INTERACTIVE DATA ANALYSIS

INTERACTIVE DATA ANALYSIS

A Practical Primer

DONALD R. McNEIL
Princeton University

A WILEY-INTERSCIENCE PUBLICATION
JOHN WILEY & SONS
New York • London • Sydney • Toronto

Library of Congress Cataloging in Publication Data:

McNeil, Donald K

Interactive data analysis.

"A Wiley-Interscience publication."

Includes index.

1. Mathematical statistics -- Data processing.

I. Title

QA276.4.M3 001.4'22 76-46571

ISBN 0-471-02631-X

Printed in the United States of America

10 9 8 7 6 5 4 3 2 1

PREFACE

Statistics, in its broadest sense, consists of three parts - exploratory data analysis, statistical inference, and stochastic model building. This book is concerned with the first part. Most statistics textbooks begin with elementary probability and then proceed to the methodology of statistical inference. In our opinion, it is more natural to begin with data analysis and end with models, for this is the order in which things usually arise in the real world, as opposed to the world of the mathematician. In scientific work, one almost always begins with the data, and when structures in the data are found to be statistically significant, a model or theory is sought to explain these patterns. John Tukey has aptly called exploratory data analysis "numerical detective work."

Despite the importance of exploratory data analysis to the development of scientific theories, little attention is paid to the subject in the statistical literature. Possible reasons for this neglect are commonly held beliefs that

(a) statistics should be concerned only with formal decision making, in which preliminary analysis of the data is disallowed because, presumably, it can influence the hypothesis and thus invalidate the testing procedure;

(b) exploratory data analysis is too simple to warrant serious discussion;

(c) exploratory data analysis is simply what a good scientist does automatically, and cannot be expounded in logical steps because of its *ad hoc* nature.

It is true that the first statement is appropriate for some applications - quality control, for example. However even if one defines statistics in its narrowest, purely inferential, sense, this is not the way things happen nowadays. Statisticians in medical centers are intimately concerned with ongoing studies which involve decision making from data (for example, which treatment should be applied to a given patient?) but which defy formal hypothesis testing because rigid alternatives are inappropriate. Economic statisticians and demographers are interested in forecasting trends and interpolating vital statistics, and here the relevance of a designed experiment is even more remote. Even in social science, where some of the most rigid practitioners of formal statistical inference are to be found, there is a growing awareness of the need for exploratory data analysis. By strictly adhering to the "hands off data" prescription, the statistician is sentenced to a role in which personal involvement in the process of scientific discovery is precluded.

We have no quarrel with statement (b): compared with the edifice of mathematical statistics, for example, exploratory data analysis _is_ simple. This fact does not, however, make it any less important. Time and again quite able statisticians have foundered in trying to fit some model to data, simply because they overlooked the possibility of a trivial transformation or a simple-minded display of residuals. The third statement is quite false. Exploratory data analysis is as logical as it is simple, as is demonstrated in the pages that follow.

To be an effective data analyst, the statistician should have certain tools, just as the pathologist needs a microscope and the surveyer requires a transit. The tools of the data analyst are a computer and a set of routine procedures. A _computer_ is a machine that can read data from some input device, manipulate these data according to a set of specific instructions, and write the result of these manipulations on some output device. The oldest computer is, of course, the human being with a pencil and a piece of paper. All of the routines we present for analyzing and displaying data can be performed by the human computer and were, in fact, designed for pencil and paper use. Since such computing can quickly become drudgery, however, we shall view the electronic computer as an extra set of hands - in the words of Richard Hamming, "an extension of the body, not the mind."

In the discussion that follows we imagine we have a computer, electronic or other, which behaves in the following way:

(a) It can read data. Since we will inevitably want to analyze different sets of data, each data array is assumed to have its own unique label. A *data array* is composed of numbers or characters in some format; the array could be a picture, for example.

(b) It can perform routine instructions. Since we will want to perform different analyses, each set of instructions is assumed to have a unique label. We will call a labeled set of instructions a *program*.

(c) It can write out the result of applying a program to a data array. The result of applying a program to a data array is another data array.

Once we have a computer that can do these things, we need a means of communicating with it, so that commands are passed that specify which program is to be applied to which data array, and what the resulting data array is to be called. If we are using an electronic computer, the usual communication devices are a terminal containing a keyboard where these commands are typed, and a terminal (often, but not necessarily, the same terminal) containing a printer or other display device on which data arrays can be written. To simplify matters, we shall assume that the commands always take the form

```
ppppp ddddd > rrrrr
```

where ppppp is the name of the program, ddddd is the name of the data array, and rrrrr is the name given to the resulting data array. We assume further that there is a standard program, "show," which can be used to write a data array on the output device. Thus to use the program ppppp on the data array ddddd and write the result out at the terminal, one would type

```
ppppp ddddd > rrrrr
show rrrrr
```

As a further simplification, we assume that these two commands can be condensed to the single command

```
ppppp ddddd
```

The assignment command > can be used to take the

result of one analysis and immediately use it as the input for another analysis. A computer that can perform in this way is called an interactive computer.

In the chapters that follow, in which we shall be presenting and using routines for exploratory data analysis, each procedure is described as it is introduced, and thereafter the discussion proceeds as if an interactive computer were available for routinely invoking the procedure. The displays that result were produced by computer programs whose components are listed at the end of the chapter in which the routine is introduced. The languages used are APL and FORTRAN.

This text originated from class notes written in September 1974, for undergraduates at Princeton University enrolled in Statistics 101 - An Introduction to Exploratory Data Analysis. The notes were intended to supplement the limited preliminary edition of John Tukey's Exploratory Data Analysis, published by Addison-Wesley, specifically, to provide the students with assistance in doing their homework problems on an interactive computer. The present text has been expanded to give more explanation of the principles of exploratory data analysis, so it could also be used as a supplement to a first course in statistical inference. Statistics 101 was inaugurated and taught for a number of years by John Tukey, and much of what has come to be called exploratory data analysis was developed by Tukey in the process of teaching this course. Our primary debt is therefore to John Tukey, whose ideas pervade almost every page of this book.

I am also indebted to my other colleagues at Princeton University, especially to Geoffrey Watson for his steadfast help and encouragement, to Ansley Coale, who patiently and benignly guided my ascent from mathematical to applied statistician, to Peter Bloomfield, whose book Fourier Analysis of Time Series: An Introduction was a timely inspiration for me, to Larry Mayer, who convinced me that people would find the material useful, and to Richard Hamming, who has been saying to me for the last year "Where is it?"

I am grateful to Susan Watkins, Michael Stoto, and Zachary Rattner, who generously gave of their time to read the first draft and suggest many improvements, and to David Donoho, whose computing and theoretical insight has been invaluable. The typing, editing, text processing, displays and final production were all done using the facilities of the Computing Laboratory in the Statistics Department at Princeton University. The software (the operating system UNIX) was provided through the generosity of the Bell Laboratories at Murray Hill, New Jersey, while the hardware (a PDP 11/40 computer) was purchased and is maintained using funds from ERDA Research Contract E (11-1)2310 awarded to the Department of Statistics. Finally, it is a pleasure to acknowledge my debt to the Office of Population Research at Princeton University which, through grants from the Ford and Rockefeller Foundations, supported my research in mathematical demography over the last five years.

Donald R. McNeil
Princeton, New Jersey
September 1976

CONTENTS

INTERACTIVE DATA ANALYSIS

CHAPTER 1

DISPLAYS

"Data! data! data!", he cried impatiently. "I can't make bricks without clay."

Sherlock Holmes, The Adventure of the Copper Beeches (Sir Arthur Conan Doyle)

WHY LOOK AT SINGLE BATCHES?

In this chapter procedures for displaying and analyzing single batches of data are described. By a single batch of numbers we mean a list of numbers that are not classified in any way. Some examples are:

(1) the average batting scores of baseball players in the American League for a given year;

(2) the areas of the major bodies of water on the earth's surface;

(3) the populations of the 50 states of the United States;

(4) the set of measurements made by a chemist on the density of a substance from repeated experiments.

Why should we be interested in such distributions of numbers? First, although most questions of scientific interest involve comparisons and relations between batches, it is important to understand the anatomy of single batches _before_ proceeding to the larger question. Before investigating in detail the relation between air pollution and health, for example, it is useful to examine the pattern of variation in these components. Again, before comparing encephalograms from patients with different heart conditions, it is wise to seek some understanding of the record for a single patient.

Second, there are occasions when a single batch of numbers is of interest in its own right. Suppose we were asked to predict the likely rainfall in Orlando next year. One way of making a prediction is to collect data on annual rainfall in Orlando for the last 100 or so years (a single batch of numbers), examine its frequency distribution, and estimate the likelihood of getting various amounts of rainfall. We might conclude, for example, that on the basis of past history it is 75% certain that the rainfall will be betwen 40 and 60 inches. (This may not be a very accurate prediction method, but it may be the best available.)

Thus there are two good reasons for knowing how to deal with a single unclassified batch of numbers - as a prelude to investigating the relation between variables, and to understand the process that generated the numbers in the batch; it is the object of this chapter to provide the techniques.

THE STEM-AND-LEAF PLOT

To begin with, we need a good way of displaying a single batch of data. Since it is assumed that the numbers are not classified in any way, their order is unimportant; therefore we might as well write them down in increasing order. This provides us with some idea of the frequency of occurrence of numbers of different magnitudes. Even better, we can produce a two-dimensional picture of the numbers, a display in which the numbers are listed line by line. The number of digits on any line corresponds to the frequency of occurrence of data between intervals of fixed length. This kind of display, called a stem-and-leaf plot, was suggested by John W. Tukey, in lectures at Princeton University (see his Exploratory Data Analysis, Addison-Wesley, 1977). In detail, the stem-and-leaf plot may be demonstrated as follows:

Consider the following data array, labeled "precipitation." These are the average amounts of precipitation, in inches, for 69 weather stations in the United States, on the basis of past records. (Source - Statistical Abstract of the United States, 1975, page 192.)

Mob.	67.0	Jun.	54.7	Phoen.	7.0	L.R.	48.5	L.A.	14.0
Sacr.	17.2	S.F.	20.7	Denv.	13.0	Hart.	43.4	Wilm.	40.2
Wash.	38.9	Jack.	54.5	Miami	59.8	Atl.	48.3	Hon.	22.9
Boise	11.5	Chic.	34.4	Peor.	35.1	Indi.	38.7	D.Mn.	30.8
Wich.	30.6	Loui.	43.1	N.Or.	56.8	Port.	40.8	Balt.	41.8
Bost.	42.5	Detr.	31.0	S.S.M	31.7	Dul.	30.2	Mn-SP	25.9
Jack.	49.2	K.C.	37.0	St.L.	35.9	G.Fls	15.0	Omaha	30.2
Reno	7.2	Conc.	36.2	At.C.	45.5	Albq.	7.8	Alb.	33.4
Buff.	36.1	N.Y.	40.2	Char.	42.7	Ral.	42.5	Bism.	16.2
Cinc.	39.0	Clev.	35.0	Col.	37.0	Ok.C.	31.4	Port.	37.6
Phil.	39.9	Pitt.	36.2	Prov.	42.8	Col.	46.4	S.Fls	24.7
Memp.	49.1	Nash.	46.0	Dall.	35.9	ElPaso	7.8	Hstn.	48.2
S.L.	15.2	Burl.	32.5	Norf.	44.7	S-Tac	38.8	Spok.	17.4
Char.	40.8	Milw.	29.1	Chey.	14.6	S.J.	59.2		

To get a stem-and-leaf plot, we first gauge the range of the data (5-70, approximately, in this case) and then, depending on the size of the array, we divide this range into intervals of fixed length.

These intervals can be units of 0.5, 1, or 2 times a power of 10. In the present case units of 5 (=0.5*10) would be acceptable. We then draw a vertical line, with the interval boundaries marked in decreasing order down its left-hand side (ignoring any decimal points or extra zeros). This is called the <u>stem</u>. In the present situation, we would arrive at Exhibit 1(a).

Next we pass through the data, in each case writing down the next significant digit on the right-hand side of the stem. These digits on the right-hand side of the display constitute the <u>leaf</u>. Proceeding through the data row by row, the value for Mobile gives rise to a 7 alongside the second 6 in the stem, since 67 is in the range 65-69; Juneau gives a 5 (54.7 rounds to 55) next to the second 5 in the stem; Phoenix yields a 7 alongside the second 0; Little Rock gives an 8 next to the second 4 (we are using the convention of rounding 5's to even numbers, so 48.5 rounds to 48, while Boise's 11.5 rounds to 12); Los Angeles gives a 7 next to the second 1, while Sacramento provides a 4 next to the first 1 in the stem. The display looks like Exhibit 1(b) after these 6 values have been inserted. After processing the remaining 63 numbers, we arrive at Exhibit 1(c).

Finally, the numbers in the leaf are sorted in increasing order on each line from left to right, yielding Exhibit 1(d).

```
 0|        0|       0|                   0|
 0|        0|7      0|7788               0|7788
 1|        1|4      1|432                1|234
 1|        1|7      1|756575             1|555677
 2|        2|       2|13                 2|13
 2|        2|       2|659                2|569
 3|        3|       3|4111200312         3|0011112234
 3|        3|       3|95976669577669     3|55666667779999
 4|        4|       4|303112032031       4|000111223333
 4|        4|8      4|889669685          4|566688899
 5|        5|       5|4                  5|4
 5|        5|5      5|579                5|579
 6|        6|       6|0                  6|0
 6|        6|7      6|7                  6|7

(a)       (b)        (c)                  (d)
```

EXHIBIT 1

A stem-and-leaf plot has a number of advantages over the more conventional histogram display. Since it contains more of the information about the batch than a histogram, it is possible to obtain reasonable estimates of the various percentage points directly from the stem-and-leaf plot.

It is also possible to use the stem-and-leaf plot to make transformations of the data and to obtain the distribution of the transformed batch of numbers, by hand without a great deal of effort.

Now let us see how we can get the computer to do the work for us. Suppose we have a program, called "stemleaf," which produces a stem-and-leaf plot. Then we just command

```
stemleaf precipitation
```

The computer responds with the following display:

```
00|7788
01|234555677
02|13569
03|00111123345566666 7789999
04|000112333333566688999
05|5579
06|07
```

This display is not the same as in Exhibit 1(d) - the computer decided to use units of 10 rather than 5 in the stem. However if we want to stretch the display, or shorten it, for that matter, we can do so by use of a scale parameter. Suppose we can do this by stating a magnification factor, which is given immediately following the command. Of course the computer will not necessarily magnify things by the exact amount we specify - it is limited to what can be done with units of 0.5, 1, and 2 times a power of 10 on the stem - but it will get as close as it can. Thus commanding

```
stemleaf precipitation {scale=2}
```

produces

```
00|7788
01|234
01|555677
02|13
02|569
03|0011112334
03|55666667789999
04|000112333333
04|566688999
05|
05|5579
06|0
06|7
```

which is very similar to Exhibit 1(d).

THE BOX PLOT

Another technique for displaying single batches of numbers, often more convenient than a stem-and-leaf plot, is the box plot, which has the advantages of simplicity and compactness. In the next chapter, where the problem of comparing different single batches is considered, the desirability of compactness becomes apparent.

A box plot is obtained by first calculating the lower and upper quartiles and the median of the batch of numbers, and then plotting these numbers on a horizontal line. The median of a batch of numbers is the value for which half of the numbers in the batch are larger and half are smaller. The lower quartile is the value that divides the batch into two parts, with 1/4 of the numbers below this value, and 3/4 above it. Similarly the upper quartile is the value for which 3/4 of the numbers are below it and 1/4 above it. In cases where these definitions give rise to a range of values (batches whose sizes are multiples of 2 or 4) the mid-point of the interval is taken. For example, in a batch consisting of the numbers 1, 2, 3, 4, 5, and 6, the median is 3.5 and the quartiles are 2 and 5, while in a batch consisting of 1, 2, 3, and 4, the median is 2.5 and the quartiles are 1.5 and 3.5.

For the 69 numbers in the array "precipitation," the median is such that 69/2 = 34.5 numbers are on either side, so the median is the 35th number in the ordered batch, which may be read off from the stem-and-leaf plot as 36. Similarly, the lower quartile is such that 69/4 = 17.25 numbers are below it, so it is the 18th number in the ordered batch, that is, 29. Finally, the upper quartile is the 18th largest number, or 43. These are plotted in Exhibit 2(a).

The next step is to draw a narrow rectangular box with ends corresponding to the lower and upper quartiles, and to replace the median point by an asterisk, yielding Exhibit 2(b). The length of this box, the <u>interquartile distance</u>, is then measured out on each side of the quartiles, giving, in the present case, marks at 15 and 57. The lowest and highest data values that fall between these quantities are also marked by crosses, joined to the box with horizontal lines, as in Exhibit 2(c). Note that in the present example there happen to be data values at 15 and 57, so the crosses are as far away from the box as they can be, but this is not always the case.

Finally, any numbers whose positions are outside these crosses are marked with circles, those more than 1.5 interquartile distances outside each quartile getting heavy circles. The result is Exhibit 2(d). Note that we have placed the digit 2 underneath two of the "outlier" positions, to indicate two values that are too close to be marked separately.

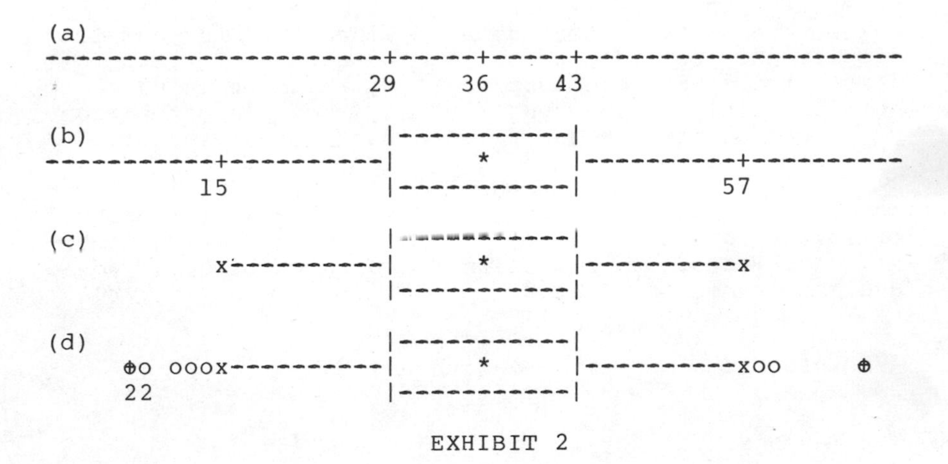

EXHIBIT 2

Thus 10 outliers are called to our special attention - the two 7's, the two 8's, 12, 13, 14, 59, 60, and 67, the first two and the last one getting heavy circles.

The box plot was suggested by John Tukey in lectures at Princeton University (see his Exploratory Data Analysis). Our definition differs slightly from that of Tukey, who puts the crosses at 1.5 interquartile distances from the quartiles. We prefer the modified definition, since in the utopian situation when the data have a Gaussian distribution, the proportion of numbers in the batch outside the crosses is, on the average, close to 1/20, a convenient fraction.

To get the computer to produce this display, we introduce a program called "boxplot," and command

```
boxplot precipitation
```

The response is

```
   7.00                                                     67.00
                   |----------|
⊕o  ooox-----------|    *     |---------x o o         ⊕
22   2             |----------|            2
```

As before, the display is slightly different from the one in Exhibit 2. To produce Exhibit 2 we calculated the quartiles from the stem-and-leaf display, rather than from the data. Since the stem-and-leaf display only records the data to an accuracy of two digits, there is some loss in information in using it to obtain quartiles. In constructing a box plot, the computer calculates the quartiles directly from the data.

Suppose that the length of the display can be changed by use of a parameter stated immediately after the command. To reduce the length from the normal setting of 50 characters to 40 characters, we would use the augmented command

```
boxplot precipitation {length=40}
```

The response is

```
   7.00                                    67.00
                |--------|
⊕o  oox---------|    *   |---------x o      ⊕
22              |--------|           2
```

So far we have been concerned simply with techniques for displaying the distribution of a single batch of numbers, without concerning ourselves with the uses to which the display may be put. Now let us begin to consider some of those uses.

TRANSFORMATIONS OF DATA

A question sometimes arises (particularly in grade school geography) as to whether Australia is a continent or an island. Some people say that it is the world's largest island, while others say that it is the smallest continent. Can this question be resolved by use of the techniques we have introduced? Let us see what can be concluded from appropriate data.

The data in this case are the land areas of all the islands and continents in the world. To include every single island, no matter how small, would yield a truly massive data array. Since the areas of the smallest islands are not particularly relevant to the question, let us concentrate on those islands which are larger than some cutoff value. There are about 50 bodies of land more than 10,000 square miles in area, and 50 is a convenient size for a batch, so let us choose 10,000 square miles as the cutoff value. The data array obtained is listed below and is given the label "islands." Asia and Europe are both included, although it could be argued that they should be combined. (source - The World Almanac and Book of Facts, 1975, page 406).

AxelH	16	Baffin	184	Banks	23	Devon	21	Elles.	82
Melv.	16	P.of W	13	South.	16	Vict.	82	Nov.Z.	32
Spits.	15	Gt.Br.	84	Irelnd	33	Green.	840	Ice.	40
Newf.	43	T d. F	19	Cuba	43	Hisp.	30	Ceylon	25
Madag.	227	Taiwan	14	Hainan	13	Hokk.	30	Honshu	89
Kyushu	14	N.Guin	306	NZ(N)	44	NZ(S)	58	Luzon	42
Mind.	36	Sakh.	29	Tas.	26	Vanc.	12	Borneo	280
Celeb.	73	Java	49	Moluc.	29	N.Br.	15	Sumat.	183
Timor	13	Asia	16988	Afr.	11506	N.Am.	9390	S.Am.	6795
Eur.	3745	Austr.	2968	Antar.	5500				

The areas are given in thousands of square miles. Requesting a stem-and-leaf plot of the data yields the following display:

```
  stemleaf islands

 ØØ|ØØØØØØØØØØØØØØØØØØØØØØØØØØØØØØØ11111222338
 Ø2|Ø7
 Ø4|5
 Ø6|8
 Ø8|4
 1Ø|5
 12|
 14|
 16|Ø
```

(In the original, the Ø in row Ø2 is underscored and the 5 in row Ø4 has a bar over it.)

(We have used the underscore to identify the position of Australia in the display.)

This display is not as revealing as it could be: most of the "islands" are bunched at one end of the display, and only the largest continent is separated. In other words, the distribution is markedly skewed. To get a more effective display, some transformation of the data is needed. Let us consider what transformation would be effective.

The problem with the above display is that the larger land masses tend to stand out: we need to reduce them in size in some way. Taking square roots of the areas would accomplish this to some extent, since the ratio of the area of the largest continent to the smallest island is 16988/1Ø = 17ØØ, while the ratio of the corresponding square roots is 13Ø/3.1 = 42. Taking logarithms might be even more successful. To transform a data array, imagine that we have a procedure called "let," which operates on a data array with the syntax

```
  let nnnnn=trans(ddddd)
```

where nnnnn is the name we choose for the transformed data array, ddddd is the name of the data array to be transformed, and trans is the transformation. To raise all the numbers in a data array to some power, we simply command

```
  let nnnnn=ddddd↑p
```

where p is the desired power. The following display shows the result of taking square roots of the island areas:

```
let z=islands↑Ø.5
stemleaf z
```

```
ØØ|3444444444444555555566666777789999944577
Ø2|9
Ø4|4
Ø6|14
Ø8|27
1Ø|7
12|Ø
```

The distribution of the square roots is only slightly more revealing than that of the raw data, since the skewness remains. Let us try taking logarithms. Suppose this can be accomplished by a function "log." We command

```
let z=log(islands)
stemleaf z
```

for which we obtain

```
Ø1|11111122222233444
Ø1|5555556666667899999
Ø2|3344
Ø2|59
Ø3|
Ø3|5678
Ø4|Ø12
```

We can now make some sense of the distribution: there appear to be two clusters, one containing the largest seven bodies of land, the other containing 41 "islands". Thus if the difference between an island and a continent is determined purely by size, there is little question that Australia is a continent, since it belongs to the cluster of larger land bodies. (Of course, some would say that this conclusion was obvious from the beginning, since Australia is not much smaller than Europe but is much larger than Greenland. But what happens if Europe and Asia were treated as a single continent? The answer would not be obvious.)

The lesson to be learned from the above example is an important one. The transformation or reexpression of data is the most powerful of the tools available to the data analyst. Used effectively, it enables one to make clearer comparisons than would otherwise be possible.

The most common motive for transforming a single batch of numbers is to arrive at a symmetric distribution of the data. Symmetry is desirable not just for aesthetic reasons. Often a "typical" or "average" value is sought to describe a single batch of numbers: differences from this typical value are regarded as experimental or random errors. It can be argued that a "typical" value summarizes a batch better if the batch is symmetric. Symbolically, we may write this decomposition as

data = typical value + residual

or, if the i-th member of the data array is x(i), as

$$x(i) = a + z(i),\ i=1,2,...,n$$

where a is the typical value and z(i) is a residual. The question of how to choose the typical value, a, which best describes a single batch, is one statisticians argue about - some prefer to use the mean, others the median, and there are other choices - but there is little controversy if the distribution is symmetric since in this case the mean and the median are the same.

Let us consider some more examples of the need for transformations. The data array "insects" is taken from Table 7 of a paper by Geoffrey Beall ("The Transformation of Data from Entomological Field Experiments," Biometrika, 1942, pages 243-262). Below are 72 numbers, these being the counts of insects in agricultural experimental units treated with different insecticides. Stem-and-leaf and box plots are also shown.

```
10  11   0   3   3  11
 7  17   1   5   5   9
20  21   7  12   3  15
14  11   2   6   5  22
14  16   3   4   3  15
12  14   1   3   6  16
10  17   2   5   1  13
23  17   1   5   1  10
17  19   3   5   3  26
20  21   0   5   2  26
14   7   1   2   6  24
13  13   4   4   4  13
```

```
  stemleaf insects

00|00111112222333333334444
00|55555556667779
01|0001112233334444
01|556677779
02|0011234
02|66

  boxplot insects

  0.00                                        26.00
        |---------------------|
x-----|          *          |---------------------x
        |---------------------|
```

Both of these displays illustrate very clearly that the insect counts, treated as a single batch, are somewhat skewed, with no outliers. However if square roots of the data are plotted, the skewness disappears:

```
  let z=insects↑0.5
  stemleaf z

00|00
00|
01|0000004444
01|77777777
02|00002222222444
02|666
03|0222333
03|556666777799
04|0011114
04|5566789
05|11

  boxplot z

  0.00                                         5.10
                   |--------------------|
x----------------|          *         |-----------x
                   |--------------------|
```

The transformation has been successful in making the distribution of the insect counts more symmetric. Note, however, the presence of two humps in the distribution, indicating the possibility that there are two kinds of experimental units, some attracting more insects than others. This bimodality was not quite so evident in the stem-and-leaf plot of the raw data, so transforming has produced an unexpected bonus.

Bimodality in a distribution raises questions - and it is precisely to raise questions about the data that we began our analysis. In the case of the islands and continents the explanation of the bimodality is obvious. In the case of the insect counts, the bimodality suggests that we need more information about the nature of the units - for example, relative positions and the insecticides used on them. We pursue this more fully in the next chapter.

Another example, given here, is the data array "rivers" containing the lengths of 141 "major" rivers in North America, as compiled by the U.S. Geological Survey. (Source - The World Almanac and Book of Facts, 1975, page 406.)

```
 735  320  325  392  524  450 1459
 135  465  600  330  336  280  315
 870  906  202  329  290 1000  600
 505 1450  840 1243  890  350  407
 286  280  525  720  390  250  327
 230  265  850  210  630  260  230
 360  730  600  306  390  420  291
 710  340  217  281  352  259  250
 470  680  570  350  300  560  900
 625  332 2348 1171 3710 2315 2533
 780  280  410  460  260  255  431
 350  760  618  338  981 1306  500
 696  605  250  411 1054  735  233
 435  490  310  460  383  375 1270
 545  445 1885  380  300  380  377
 425  276  210  800  420  350  360
 538 1100 1205  314  237  610  360
 540 1038  424  310  300  444  301
 268  620  215  652  900  525  246
 360  529  500  720  270  430  671
1770
```

The displays of the distribution are obtained using the

following commands:

```
  stemleaf rivers


 ØØ|4
 Ø2|Ø11223334555566667778888899990000111122333334445+16
 Ø4|11122233344556677900123334456 7
 Ø6|ØØØ11223357801223446 8
 Ø8|Ø45790018
 1Ø|Ø45Ø7
 12|1471
 14|56
 16|7
 18|9
 2Ø|
 22|25
 24|3
 26|
 28|
 3Ø|
 32|
 34|
 36|1
```

```
  boxplot rivers


 135.ØØ                                               371Ø.ØØ
  |----|
x-| *  |-----xooΦ Φ    Φ Φ      Φ  Φ                 Φ
  |----|       22 2             2
```

(For printing convenience, the second line in the stem-and-leaf display is truncated if any leaf contains more than 5Ø digits. This plot contains 16 numbers in the second line of the leaf that are not printed.)

After some trial and error (trying square roots, logarithms, and reciprocals), the reciprocal square root transformation of the river lengths seems to give the most symmetric distribution:

```
  let z=rivers↑-Ø.5
  stemleaf z
```

```
Ø1|6
Ø2|Ø1134
Ø2|6688899
Ø3|Ø1122333444
Ø3|55667777788899
Ø4|ØØØØØ111122333334444
Ø4|5556677777888999999
Ø5|Ø111111223333333344
Ø5|5555556667778888999
Ø6|ØØØØØ1112223333 4
Ø6|56668899
Ø7|Ø
Ø7|
Ø8|
Ø8|6
```

boxplot z

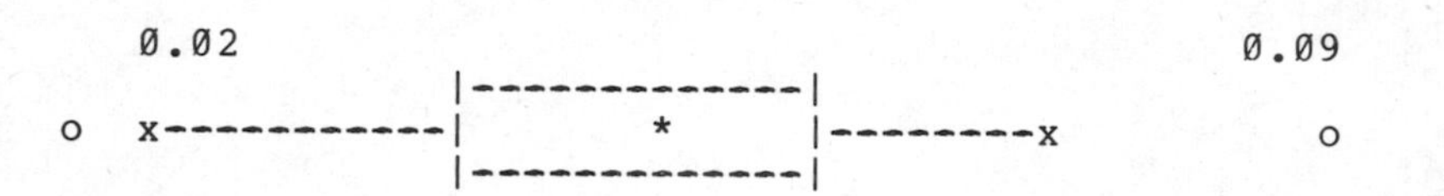

Why should reciprocal square roots be a more useful way of expressing river lengths? We offer no explanation but hope that geographers may contribute one.

NUMERICAL SUMMARIES

The representation of the data as the sum of a typical value and a residual leads to a useful way of summarizing a batch of numbers. In many situations, if the distribution is reasonably symmetric, knowing the size of the errors is all that matters. In this case we may be happy to replace the batch by two or three numbers - the typical value, a measure of the size of the residuals, and, possibly, the number of elements in the data array.

In the box plot display the interquartile distance was used as a measure of spread, and we shall use this as our measure of size in residuals. On occasion, it is useful to augment this summary by two additional numbers - the two extreme values of the batch.

Suppose a routine is available for providing these summaries. Call it "condense," and let it produce as output the median and the interquartile distance (the <u>midspread</u>, for short) of the batch. To produce a five-number summary consisting of the sample size, the minimum, the lower quartile, the median, the upper quartile, and the maximum, suppose we can change a parameter called "num." If the parameter is 5, a five-number summary is produced. If it is 7, a summary consisting of the above five statistics together with the 12.5% and 87.5% percentiles (the "eighths") is produced, while 1 produces medians only and 2 produces medians and midspreads, the default option. For example the command

```
  condense rivers
```

yields

```
  425.000  370.000
   med      sprd
```

while

```
  condense rivers {num=5}
```

gets the response

```
135.000   310.000   425.000   680.000  3710.000   141
 min       loq       med       upq      max      size
```

EXERCISES

1. The ages at death of the rulers of the United King dom, starting with William the Conqueror, are as follows (source - The World Almanac and Book of Facts, 1975, page 357):

```
Will1 60 Will2 43 Hen1  67 Steph 50 Hen2  56 Rich1 42
John  50 Hen3  65 Edw1  68 Edw2  43 Edw3  65 Rich2 34
Hen4  47 Hen5  34 Hen6  49 Edw4  41 Edw5  13 Rich3 35
Hen7  53 Hen8  56 Edw6  16 Mary1 43 Eliza 69 Jmes1 59
Chas1 48 Crom1 59 Crom2 86 Chas2 55 Jmes2 68 Will3 51
Mary2 33 Anne  49 Geo1  67 Geo2  77 Geo3  81 Geo4  67
Will4 71 Vict  81 Edw7  68 Geo5  70 Edw8  77 Geo6  56
```

Make a stem-and-leaf display of these data. Compute the median and the upper and lower quartiles of the distribution. Hence find an interval that contains the lifetime of a randomly selected ruler from the above population with 50% probability.

2. The number of fatalities in some notable aircraft crashes since 1937 are given as follows (source - The World Almanac and Book of Facts, 1975, page 817):

May37	36	Aug44	76	Jul45	14	Nov49	55	Jun50	58
Dec51	56	Dec52	87	Mar53	11	Jun53	129	Nov55	44
Jun56	74	Jun56	128	Aug57	79	Feb59	65	Feb60	61
Mar60	63	Jul60	13	Dec60	134	Feb61	73	Sep61	78
Sep61	83	Nov61	77	Mar62	95	Mar62	111	Mar62	107
Jun62	130	Jun62	113	Nov62	97	Feb63	95	Jun63	101
Nov63	118	Dec63	82	Feb64	83	Mar64	85	May64	75
Feb65	84	May65	121	Jan66	117	Feb66	133	Mar66	124
Apr66	82	Sep66	97	Dec66	129	Mar67	26	Apr67	126
Jun67	88	Jun67	72	Jul67	82	Oct67	66	Nov67	68
Dec67	66	Apr68	122	May68	85	Sep68	95	Mar69	155
Mar69	87	Jun69	79	Sep69	83	Nov69	87	Dec69	93
Feb70	102	Jul70	112	Jul70	108	Aug70	101	Oct70	30
Nov70	75	Dec70	90	May71	78	Jul71	162	Aug71	97
Sep71	111	Mar72	112	Aug72	156	Oct72	176	Dec72	155
Dec72	100	Jan73	176	Apr73	104	Jun73	14	Jul73	122
Jul73	68	Jul73	89	Aug73	85				

Make a stem and leaf display of these data and comment on its shape. Compute the five-number summary.

3. The mean annual temperature in degrees Fahrenheit in New Haven, Connecticut, from 1912 to 1971 is as follows (source - an article by J.E. Vaux and N.B. Brinker, <u>Cycles</u>, 1972, pages 117-121). Make a pictorial display of this distribution. What is the chance that the mean temperature in a given year in future will exceed 52 degrees, based on these data?

49.9	52.3	49.4	51.1	49.4	47.9	49.8	50.9
49.3	51.9	50.8	49.6	49.3	50.6	48.4	50.7
50.9	50.6	51.5	52.8	51.8	51.1	49.8	50.2
50.4	51.6	51.8	50.9	48.8	51.7	51.0	50.6
51.7	51.5	52.1	51.3	51.0	54.0	51.4	52.7
53.1	54.6	52.0	52.0	50.9	52.6	50.2	52.6
51.6	51.9	50.5	50.9	51.7	51.4	51.7	50.8
51.9	51.8	51.9	53.0				

4. The following data comprise the gestation in days and the longevity in years for various mammals (source - The World Almanac and Book of Facts, 1975, page 463):

ass	365	24	baboon	187	27	badger	60	15
bl.bear	219	19	gr.bear	225	31	pl.bear	240	31
beaver	122	13	buffalo	278	20	camel	406	20
dom.cat	63	15	chimp.	231	30	chipmunk	31	7
cow	284	18	deer	201	17	dog	61	16
elephnt	645	47	elk	250	22	fox	52	8
giraffe	425	10	mt.goat	184	9	gorilla	257	25
guin.pig	68	4	horse	330	27	kangaroo	42	19
leopard	98	17	lion	100	22	monkey	164	7
moose	240	8	mouse	21	4	opossum	16	4
pig	112	14	puma	90	11	rabbit	37	5
rhinoc.	450	27	s.lion	350	19	sheep	154	13
squirrel	44	8	tiger	105	19	whale	365	37
wolf	63	12	zebra	365	20			

Make stem-and-leaf displays and box plots for each distribution separately. In each case would the distribution be more symmetric if some reexpression were used? If so, make displays of the transformed distributions.

5. According to the 1970 United States census of population, the populations of the 50 states were (in millions)

Ala	3.44	Alaska	0.30	Ariz	1.77	Ark	1.92	Cal.	19.95
Colo	2.21	Conn	3.03	Del	0.55	Fla	6.79	Ga	4.59
Haw.	0.77	Idaho	0.71	Ill	11.11	Ind	5.19	Iowa	2.83
Kan	2.25	Ky	3.22	La	3.64	Me	0.99	Md	3.92
Mass	5.69	Mich	8.88	Minn	3.81	Miss	2.22	Mo	4.68
Mont	0.69	Neb	1.48	Nev	0.49	N.H.	0.74	N.J.	7.17
N.M.	1.02	N.Y.	18.24	N.C.	5.08	N.D.	0.62	Ohio	10.65
Okla	2.56	Ore	2.09	Pa	11.79	R.I.	0.95	S.C.	2.59
S.D.	0.67	Tenn	3.92	Texas	11.20	Utah	1.06	Vt	0.44
Va	4.65	Wash	3.41	W.Va	1.74	Wis	4.42	Wyo	0.33

Make a stem-and-leaf plot of these data and comment on its appearance. What transformation produces a reasonably symmetric distribution? Make a stem-and-leaf plot of the transformed data. What conclusions can be made?

6. The estimated diameters in miles of some bodies in the solar system are listed (source - Encyclopaedic Dictionary of Physics, Vol 6, 1962 (J. Thewlis, ed.), pages 533-534). By using an appropriate display find a suitable reexpression for analyzing these data. Comment on the shape of the distribution, mentioning any particular features of interest.

Sun	864000	Merc.	3100	Venus	7700	Earth	7914
Moon	2160	Mars	4200	Phobos	10	Demos	5
Jupit.	85750	Amal.	100	Io	2000	Europa	1750
Gany.	3000	Call.	2800	Hestia	80	Hera	25
Demet.	12	Adast.	12	Pan	15	Poseid	25
Hades	12	Saturn	71150	Mimas	300	Encel.	400
Tethys	600	Dione	600	Rhea	850	Titan	3000
Hyper.	250	Iapet.	750	Phoebe	150	Uranus	29300
Miran.	100	Ariel	400	Umbr.	300	Titan.	600
Oberon	500	Nept.	27700	Triton	2500	Nereid	200
Pluto	4900						

7. Collect a single batch of data that interests you, and perform an analysis according to the techniques outlined in Chapter 1.

PROGRAM COMPONENTS

The following program components are supplied for the benefit of electronic computer enthusiasts. For many applications involving small to moderate data arrays, APL is an ideal language, and the following functions are sufficiently concise to be entered into the computer by hand. FORTRAN subroutines are also given. While FORTRAN is more efficient than APL for repeated analyses on large arrays, it requires system-dependent programs for reading and writing data arrays, and the code is somewhat longer than that for APL. Some changes may also be necessary in the FORTRAN code to adapt it to local conditions. (The FORTRAN programs were developed on a PDP 11 system, using a Princeton University adaptation of RT-11 FORTRAN software.)

In the interest of conciseness, the programs are listed without interspersed comments. Brief explanations of the algorithm, where necessary, and the function of parameters are given before each listing. While the programs have been tested on the data arrays in this book, they cannot be guaranteed to be completely error free.

Programs for the stem-and-leaf display and the box plot are given in this section. The algorithm for the stem-and-leaf display uses a naive but serviceable scaling routine based on the the range and size of the data array. There are three parameters that can be changed by the user - "scale" (mentioned in the text), which varies the depth of the stem-and-leaf display and should default to the value 1.0, "width" ("iwidth" in FORTRAN), which controls the width to which the display is truncated (up to 160), and "atom," which is added to the range of the data to avoid a zero division if the range is less than the smallest nonzero number the computer can process. The default for "width" should depend on the number of characters that the terminal can print on a single line, while that for "atom" depends on the precision of the computer.

The procedure for the box plot involves use of a subprogram, "fill," which does the necessary computations to determine where the characters should be placed in the display. As mentioned in the text, the parameter "length" controls the horizontal dimension of the display.

APL FUNCTIONS

All of the APL functions listed in this book use a single right-hand argument, with the exception of the auxiliary function "FILL." In contrast to the computing syntax used in the text, APL functions operate from right to left. For example, to store the output of the function "FFFFF" applied to the array "X" in the file "Z," one would use the command

```
Z < FFFFF X
```

```
      ∇STEMLEAF[⎕]∇
      ∇ Z←STEMLEAF X;C;R;S;SI;I;J;F;A;W;L
[1]   C←10*1-⌊10⍟R←ATOM+(X[ρX]-(X←X[⍋X←,X])[1])÷SCALE
[2]   SI←(1+3×+/(R×C)> 25 50 )++/(ρX)> 25 100
[3]   X←⌊0.5+X×C×10*+/SI= 2 3 6
[4]   F←+/(SI=ι9)/ 0.5 2 1 1 0.5 2 2 1 0.5
[5]   A←20ρ '01234567890123456789' ,Z←''
[6]   I←F×⌊X[1]÷10×F
[7]   S← '- ' [1+J←X[1]≥0]
[8]   L←1+(|W←(X≤(10×I)+J×¯1+10×F)/X)-10×⌊|I
[9]   Z←Z,WIDTHρS,A[1+( 10 10 )⊤|⌊I],'|',A[L],WIDTHρ' '
[10]  I←I+F×1-(I=0)×X[1]<0
[11]  →7×ι0<ρX←(ρW)↓X
[12]  Z←((⌊(ρZ)÷WIDTH),WIDTH)ρZ
```

```
      ∇BOXPLOT[⎕]∇
      ∇ Z←BOXPLOT X;L;U;B;W
[1]     W←LENGTH
[2]     X←(÷ATOM+U-L)×((U←⌈/,X)-W×L←⌊/,X)+,X×W-1
[3]     (⍕0.1×⌊0.5+10×L),(⍕(0⌈W-6)⍴' '),⍕0.1×⌊0.5+10×U
[4]     Z←W FILL X
[5]     B← ' ¯|⌈||¯⎕                          ' [1+Z]
[6]     B←B, ' *|*|* *-× ooooooooo ⊛⊛⊛⊛⊛⊛⊛⊛⊛' [1+Z]
[7]     B←B, ' _|⌊||_⎕    23456789  23456789' [1+Z]
[8]     Z←(3,W)⍴B
      ∇


      ∇FILL[⎕]∇
      ∇ Z←W FILL X;N;D;X1;X2;X3;X4;Y;I;J;S
[1]     N←⍴X←X[⍋X←,X]
[2]     Z←0.5×X[⌊0.25×3+N×⍳3]+X[⌈0.25×1+N×⍳3]
[3]     S←(Z[1]-0.5+1.5×D),(Z[1]-0.5+D←Z[3]-Z[1]),Z[1]
[4]     S←⌊0.5+S,Z[2],Z[3],(Z[3]+0.5+D),Z[3]+0.5+1.5×D
[5]     X1←(X2≤S[1])/X2←(X≤S[2])/X←⌊0.5+X
[6]     X2←(X2>S[1])/X2
[7]     X4←(X3≥S[7])/X3←(X≥S[6])/X
[8]     X3←(X3<S[7])/X3
[9]     Y←((X>S[2])×X<S[6])/X
[10]    Z←W⍴0
[11]    Z[I]←20+9⌊+⌿(¯1,X1)∘.=I←⍳(S←0⌈S⌊W)[1]
[12]    Z[I]←10+9⌊+⌿(¯1,X2)∘.=I←S[1]+⍳1+S[2]-S[1]
[13]    Z[I]←10+9⌊+⌿(¯1,X3)∘.=I←¯1+S[6]+⍳1+S[7]-S[6]
[14]    Z[I]←20+9⌊+⌿X4∘.=I←¯1+S[7]+⍳1+W-S[7]
[15]    →18×⍳0=⍴Y
[16]    Z[J←¯1↑Y]←Z[I←1↑Y]←9
[17]    Z[I+⍳0⌈J-I+1]←8
[18]    Z[S[3]+⍳0⌈S[5]-S[3]+1]←6
[19]    Z[S[3],S[5]]←0
[20]    Z[S[4]]←1
[21]    Z[S[3],S[5]]←Z[S[3],S[5]]+ 2 4
      ∇
```

FORTRAN SUBROUTINES

Since the sort operation is not a library function in FORTRAN as it is in APL, a subroutine for sorting a single list of numbers (Shell sort) is given below, in addition to the other programs. For the subroutines "stemleaf" and "boxplot," the data are assumed to be stored in a one-dimensional array, x, of cize n.

```
      subroutine stemleaf(x,n,scale,iwidth,atom)
      dimension x(n),lf(5ØØ),ia(2Ø),m(4)
      data ia/'Ø','1','2','3','4','5','6','7','8','9',
     +'Ø','1','2','3','4','5','6','7','8','9'/
      data m/'-',' ','|','+'/
      call sort(x,n)
      r=atom+(x(n)-x(1))/scale
      c=1Ø.**(11-int(alog1Ø(r)+1Ø))
      mm=minØ(2,maxØ(int(r*c/25.),Ø))
      k=3*mm+2-15Ø/(n+5Ø)
      if((k-1)*(k-2)*(k-5).eq.Ø) c=c*1Ø
      mu=1Ø
      if(k*(k-4)*(k-8).eq.Ø) mu=5
      if((k-1)*(k-5)*(k-6).eq.Ø) mu=2Ø
      i=1
      if(x(1).ge.Ø) i=2
      ii=1
      d=mu*(int(x(ii)*c/mu)+i-2)/1Ø.
1     do 2 k=1,iwidth
2     lf(k)=m(2)
      if(i.eq.2.or.d.le.Ø) goto 3
      i=2
      d=d-mu/1Ø.Ø
3     j=Ø
4     j=j+1
      ix=int(Ø.5+abs(x(ii)*c-1Ø*int(d)))
      if((x(ii)*c-1Ø*d).ge.Ø.5+(mu-1)*(i-1)) goto 5
      if(j.le.iwidth) lf(j)=ia(1+ix)
      ii=ii+1
      if(ii.gt.n) goto 5
      goto 4
5     id=mod(iabs(int(d)),1ØØ)
      k1=1+id/1Ø
      k2=1+id-1Ø*(k1-1)
      if(j.le.iwidth+1) goto 6
      lf(iwidth-2)=m(4)
      lf(iwidth-1)=ia(1+(j-iwidth+2)/1Ø)
      lf(iwidth)=ia(j-iwidth+3-1Ø*((j-iwidth+2)/1Ø))
6     k=minØ(iwidth,j)
      write(6,7) m(i),ia(k1),ia(k2),m(3),(lf(j),j=1,k)
7     format(132a1)
      if(ii.gt.n) return
      d=d+mu/1Ø.Ø
      goto 1
      end
```

```
         subroutine fill(x,n,iy,isize)
         dimension x(n),iy(isize)
         zl=Ø.5*(x(int(Ø.25*n+Ø.75))+x(int(Ø.25*n+1.)))
         zm=Ø.5*(x(int(Ø.5*n+Ø.75))+x(int(Ø.5*n+1.)))
         zu=Ø.5*(x(int(Ø.75*n+Ø.75))+x(int(Ø.75*n+1.)))
         d=zu-zl
         il=maxØ(int(Ø.5+zl-1.5*d),1)
         i2=maxØ(int(Ø.5+zl-d),1)
         i3=int(Ø.5+zl)
         i4=int(Ø.5+zm)
         i5=int(Ø.5+zu)
         i6=minØ(int(Ø.5+zu+d),isize)
         i7=minØ(int(Ø.5+zu+1.5*d),isize)
         do 1 k=1,isize
1        iy(k)=2Ø
         do 2 k=il,i7
2        iy(k)=1Ø
         do 3 k=i2,i6
3        iy(k)=Ø
         do 4 j=1,n
         ie=int(Ø.5+x(j))
         if(ie.ge.il) goto 5
4        iy(ie)=iy(ie)+1
5        do 6 k=1,n
         ie=int(Ø.5+x(n+1-k))
         if(ie.le.i7) goto 7
6        iy(ie)=iy(ie)+1
7        do 8 l=j,n
         ie=int(Ø.5+x(l))
         if(ie.ge.i2) goto 9
8        iy(ie)=iy(ie)+1
9        do 1Ø m=k,n
         ig=int(Ø.5+x(n+1-m))
         if(ig.le.i6) goto 11
1Ø       iy(ig)=iy(ig)+1
11       do 12 j=ie,ig
12       iy(j)=8
         iy(ie)=9
         iy(ig)=9
         do 13 j=i3,i5
13       iy(j)=6
         iy(i3)=Ø
         iy(i5)=Ø
```

```
      iy(i4)=1
      iy(i3)=iy(i3)+2
      iy(i5)=iy(i5)+4
      do 14 i=i1,i7
 14   iy(i)=min0(iy(i),19)
      do 15 i=1,isize
 15   iy(i)=min0(iy(i),29)
      return
      end



      subroutine boxplot(x,n,length,atom)
      dimension x(n),iy(160),iz(3,160)
      dimension frmat(16),ia(30),ib(30),ic(30)
      logical*1 frmat
      data ia/' ','-','|','|','|','|','-','x',' ',' ',
     +' ',' ',' ',' ',' ',' ',' ',' ',' ',' ',' ',
     +' ',' ',' ',' ',' ',' ',' ',' ',' '/
      data ib/' ','*','|','*','|','*',' ','*','-','x',
     +' ','o','o','o','o','o','o','o','o','o',' ',
     +'@','@','@','@','@','@','@','@','@'/
      data ic/' ','-','|','|','|','|','-','x',' ',' ',
     +' ',' ','2','3','4','5','6','7','8','9',' ',
     +' ','2','3','4','5','6','7','8','9'/
      call sort(x,n)
      encode(16,1,frmat) length-14
1     format('(f7.2,',i3,'x,f7.2)')
      write(6,frmat) x(1),x(n)
      a=(length-1.)/(atom+x(n)-x(1))
      b=1.-a*x(1)
      do 2 i=1,n
2     x(i)=a*x(i)+b
      call fill(x,n,iy,length)
      do 3 i=1,length
      iz(1,i)=ia(1+iy(i))
      iz(2,i)=ib(1+iy(i))
3     iz(3,i)=ic(1+iy(i))
      do 4 j=1,3
4     write(6,5) (iz(j,i),i=1,length)
5     format(160a1)
      return
      end
```

```
      subroutine sort(x,n)
      dimension x(n)
      i=1
1     i=i+i
      if(i.le.n) goto 1
      m=i-1
2     m=m/2
      if(m.eq.Ø) return
      k=n-m
      do 4 j=1,k
      kk=j
3     if(kk.lt.1) goto 4
      if(x(kk+m).ge.x(kk)) goto 4
      w=x(kk+m)
      x(kk+m)=x(kk)
      x(kk)=w
      kk=kk-m
      goto 3
4     continue
      goto 2
      end
```

CHAPTER 2

COMPARISONS

A picture is worth more than ten thousand words
Chinese Proverb

SCHEMATIC PLOTS

In the preceding chapter we introduced methods for effectively displaying a single batch of numbers. In practice we need to be able to analyze more diverse data arrays, data classified in different ways. Some of the exercises at the end of Chapter 1 suggest some kind of comparison or relation, that is obscured by regarding the data as a single batch of numbers. For the first two examples (British rulers' longevities and air disaster fatalities, respectively) one might look for time trends in the data. For the mammal data (Exercise 5) one could look for some relation between the gestation period and the longevity. In the case of the solar system body diameters, it might be useful to classify the bodies into planets and moons and to compare the two distributions.

In data analysis, as in most detective work, it is important to begin with a general picture - a view from the height - and then to focus more and more sharply on the features of interest. The purpose of the stem-and-leaf plot is to give this overall view. In this chapter we begin to look at the details below the surface. The first step in this search involves comparing batches.

A typical practical problem may be described as follows. Yarn of six different types is produced at a factory. It is desired to evaluate the strengths of the various types. Measurements are made on the number of warp breaks per loom, where a loom corresponds to a fixed length of yarn. When 54 looms (9 from each type) are examined, the results are as follows (source - L.H.C. Tippett, Technological Applications of Statistics, Wiley, 1950, page 106):

Type 1	2	3	4	5	6
26	18	36	27	42	20
30	21	21	14	26	21
54	29	24	29	19	24
25	17	18	19	16	17
70	12	10	29	39	13
52	18	43	31	28	15
51	35	28	41	21	15
26	30	15	20	39	16
67	36	26	44	29	28

How can one analyze these data?

To begin with, we need an effective display. It would be desirable to reduce each of the six batches to a one-dimensional display, and to put these figures next to each other. Box plots are ideal for this purpose, since we can immediately compare the centers and spreads of the distributions of the batches and spot outliers. Suppose we have a routine, called "compare," which produces vertical boxplots parallel to each other, and writes out the minimum and the maximum value of all the data in the array. To vary the number of lines in the display, a parameter "depth" is used whose default value is 20. In the present example we would invoke the routine using the command

```
compare warpbreaks
```

and obtain

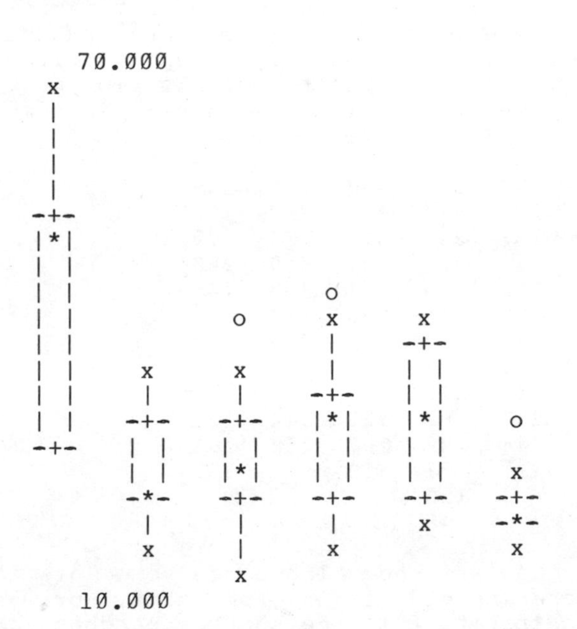

This display of the centers and spreads of the distributions of warp breaks for each of the six types of yarn makes it immediately apparent that the looms in the first batch have a larger number of warp breaks, on average, than those in the other batches. If a numerical summary is desired, we can use the program "condense" to produce the values of the midspreads and medians of the batches in the various columns:

```
  condense warpbreaks

    51.000   28.000
    21.000   12.000
    24.000   10.000
    29.000   11.000
    28.000   18.000
    17.000    6.000
     med     sprd
```

We can also use box plots to compare batches of unequal size. An instance might arise as follows: An experiment was conducted to measure and compare the effectiveness of various diet supplements on the growth rate of chickens. Newly hatched chicks were randomly allocated into six groups, and each group was given a different feed supplement. Their weights in grams, after six weeks, were as follows (source - Anonymous query, <u>Biometrics</u>, 1948, page 214):

179	309	243	423	325	368
160	229	230	340	257	390
136	181	248	392	303	379
227	141	327	339	315	260
217	260	329	341	380	404
168	203	250	226	153	318
108	148	193	320	263	352
124	169	271	295	242	359
143	213	316	334	206	216
140	257	267	322	344	222
	244	199	297	258	283
	271	171	318		332
		158			
		248			

The groups are unequal because some of the chicks died before the end of the experiment. For ease of storage it is better to fill the shorter batches with "missing values" to make them all the same size, so that the data array is a rectangular array. However we need to tell our computer to ignore the missing values when summarizing the distribution. Assume that this can be done by resetting a parameter, "miss," when the program command is given. If the data array is called "chickwts" and the missing values are set to zeros, schematic plots and five-number summaries are obtained by the commands which follow, the results being given in each case.

compare chickwts {miss=Ø}

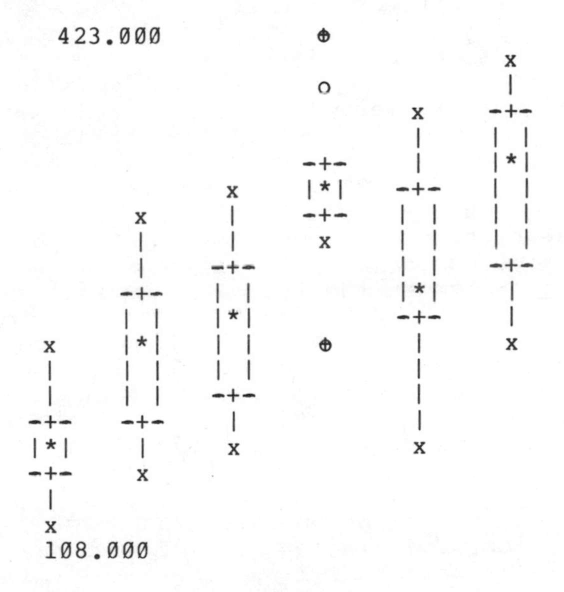

condense chickwts {num=5;miss=Ø}

min	loq	med	upq	max	size
1Ø8.ØØØ	136.ØØØ	151.5ØØ	179.ØØØ	227.ØØØ	1Ø
141.ØØØ	175.ØØØ	221.ØØØ	258.5ØØ	3Ø9.ØØØ	12
158.ØØØ	199.ØØØ	245.5ØØ	267.ØØØ	329.ØØØ	14
226.ØØØ	3Ø7.5ØØ	328.ØØØ	34Ø.5ØØ	423.ØØØ	12
153.ØØØ	242.ØØØ	263.ØØØ	325.ØØØ	38Ø.ØØØ	11
216.ØØØ	271.5ØØ	342.ØØØ	373.5ØØ	4Ø4.ØØØ	12

The schematic comparison display shows the differences between the effects of the diet supplements quite clearly, as well as pinpointing some possible outliers in the fourth batch. Apart from the fourth batch, there is a steady pattern of increasing weight gains as we move across the batches.

TRANSFORMATIONS OF SEVERAL BATCHES

In the preceding chapter we argued that it was more desirable to obtain a symmetric distribution of the data which could be accomplished through a transformation. When dealing with several batches it might be reasonable to seek a transformation that makes each as symmetric as possible, but this is hard to do, and, in any case, there is a situation more desirable than symmetry when batches are being compared.

Remember that we are interested in comparing the typical values of the numbers in the various batches. This comparison is easy if the spreads are all small in relation to the differences between the typical values, since most of the numbers in each batch are then close to their typical value, and the typical values can be said to represent the batches. If the spreads are all large, the comparison may still be easy - we may be forced to conclude simply that there are no detectable differences between the typical values of the batches. But if some of the spreads are small and others large, it may be difficult to make a judgment, since this means that some typical values can be obtained more accurately than others - a kind of confusion that makes comparison difficult. In other words, the comparison of the levels of the typical values is obscured by the spread.

Just as making an appropriate transformation can improve our vision of a single batch of numbers, a transformation can increase our ability to make valid comparisons between batches. In this case we are guided primarily not by a desire for symmetry but by a desire to even out the spreads. It may not be possible to accomplish this to our satisfaction, but we can at least remove a relation between the spreads and the centers of the batches. A convenient vehicle for displaying such relations is a scatter plot.

To get a scatter plot of a batch of (x,y) values, two lines at right angles to each other are drawn, one, the x-axis, horizontal, the other, the y-axis, vertical. The intersection of these straight lines is called the origin. Each (x,y) pair is marked as a cross (or asterisk or small circle or other symbol). The position of the point whose values are x and y is obtained by measuring a distance x along the x-axis from the origin (to the right if x is positive, to the left if x is negative) and then measuring a distance y up (or down, if y is negative). Suppose we have a routine for doing this, called "scat," which acts on a data array consisting of two columns, with the values of x in the first column, and the values of y in the second column. The resulting display is a scatter plot.

In the present case, we need a plot of the midspreads of the batches against their medians. This may be accomplished by first using the routine "condense" to get the medians and the midspreads, storing the output from this program under some temporary label, and then using the temporary label as the argument to the routine "scat." For the data array "warpbreaks," we may proceed as follows:

```
condense warpbreaks > z
scat z
```

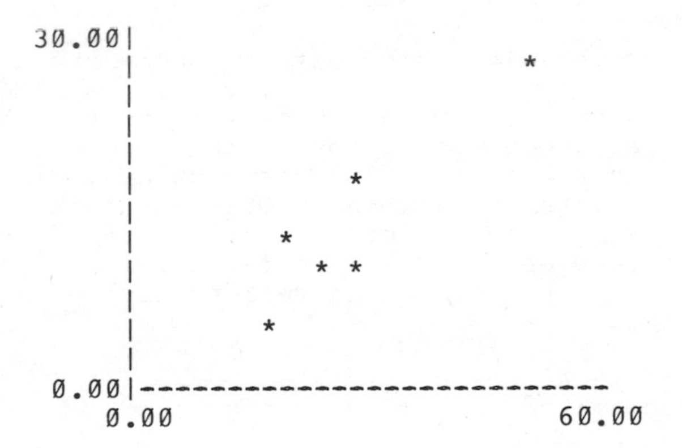

It is clear from the preceding display that there is a positive relationship between the midspreads and the medians of the warp break counts. If a straight line were fitted to the points in the graph, its slope would be about 0.5, since there is an increase of 30 in the vertical coordinate corresponding to an increase of 60 in the horizontal coordinate, and the ratio of 30 to 60 is 0.5. This is interesting information in itself, but the comparison of the batches is still obscured. The comparison would be easier if there were no relation between the midspreads and the medians, so that we would be comparing only the medians. To even out the spreads, we need to pull down the higher values, which will reduce the spread and facilitate comparison. Let us try taking logarithms of the counts, and then observing the relation between the spreads and centers of batches of the transformed data:

```
let logs=log(warpbreaks)
condense logs > z
scat z
```

```
Ø.32|                   *
    |
    |
    |         *
    |
    |   *
    |      *  *
    |
    |*
Ø.12|-------------------------
   1.2Ø                    1.8Ø
```

Notice that in this display the axes no longer intersect at the origin. In some cases more resolution is obtained by moving these lines so that they intersect at some point other than the origin. Assume that the program "scat" produces such a modified scatter plot, where desirable. Observe that the transformation has reduced but not removed the dependence between spreads and centers; there is still a pattern remaining. Let us try taking reciprocals of the counts:

```
let recips=warpbreaks↑-1
condense recips > z
scat z
```

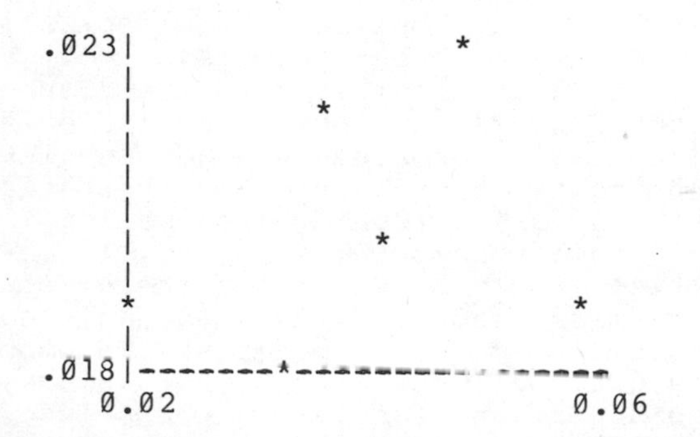

There is no detectable relationship left between the spreads and the centers. To see how even the spreads of the transformed data are, we can print out the file "z" containing the medians and the spreads:

```
show z

   0.020    0.019
   0.048    0.023
   0.042    0.020
   0.034    0.018
   0.036    0.022
   0.059    0.019
  med      sprd
```

Let us now go back and look at a comparison display of the reciprocals of the data, by using the program "compare." We obtain, stretching the display slightly to give more detail,

```
compare recips {depth=25}
```

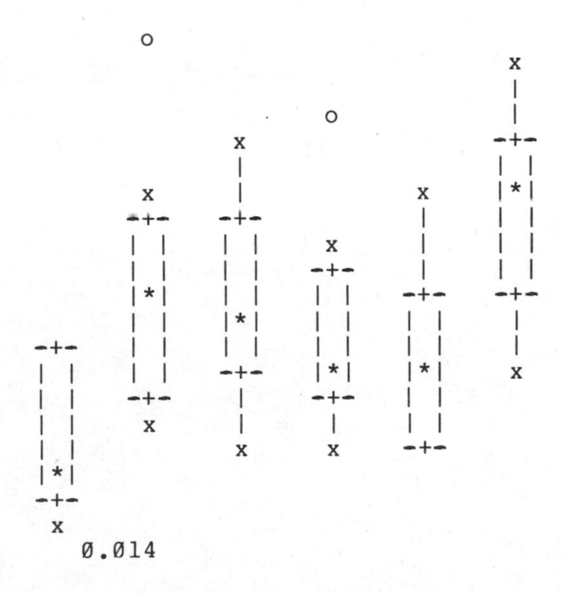

Although the conclusion obtained earlier is not altered - the yarn in the first batch seems to be less resistant to breaks than that in the other batches - we have learned something more as a result of the transformation. There are a couple of outliers which may be worth further examination, and there may be some differences between the other yarn types. More importantly, the success of our transformation in removing pattern suggests that comparison of batches may be facilitated by a change of scale. In this case, using the reciprocal changed the scale by which we measured yarn strength from breaks/unit length to average distance between breaks. The typical value now better represents the strength of each batch.

As a further example of the need for transformation when comparing batches, let us return to the insect counts considered in Chapter 1. In fact (something not mentioned when the insect data were introduced), the data array is not a single batch of 72 numbers, but six batches of 12 numbers, with each column corresponding to a separate batch. The experimental units in each batch are treated with different insecticides. A schematic plot of their distributions is obtained as follows:

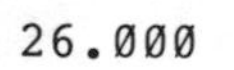

compare insects

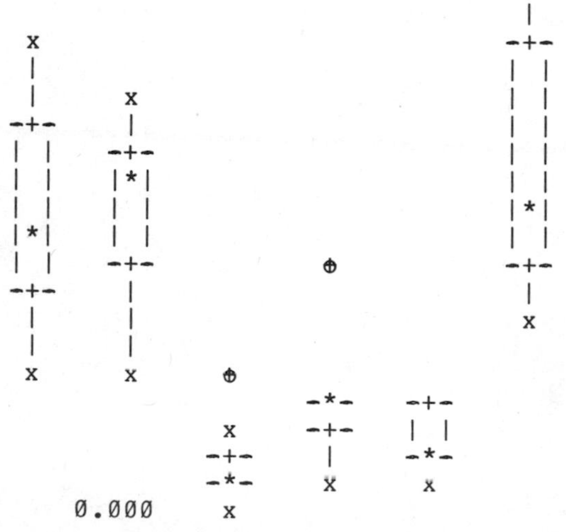

It is obvious from this display that three of the batches have more insects on average than the other three. This is the cause of the slight but perceptible bimodality that was evident in the stem-and-leaf plot of the square roots of the counts, shown in Section 4 of Chapter 1.

When the batch as a whole was considered, square roots were successful in removing skewness. In the present case, however, there is a question as to whether taking square roots completely removes the relationship between the midspreads and medians of the batches, as can be seen from the following analysis:

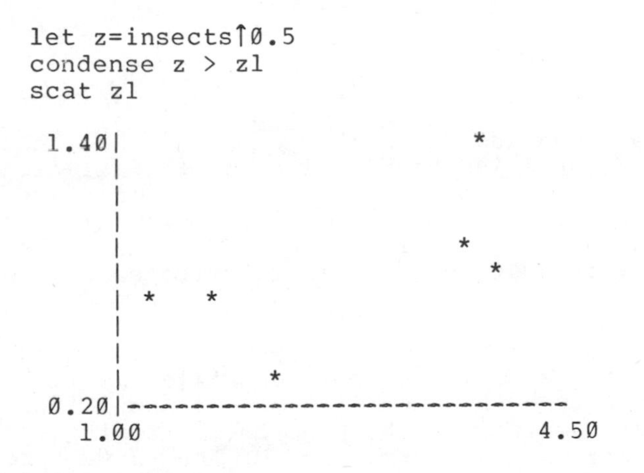

In the analysis of the island-continent data we found that logarithms were more effective than square roots in removing skewness from the distribution of a batch. Having had this experience, we might consider taking logarithms of the insect counts, in the hope that the relation in the preceding plot would be removed. However there is a problem with taking logarithms of the insect counts: how does one handle the zeros?

The simplest and most effective way of making logarithm transformations of data containing zeros is to add a small constant amount to each number before making the transformation. Ideally, the value to add could be estimated from the data in some way, but when dealing with counts a value between 0.1 and 1.0 is usually satisfactory. In the present case, we choose to add unity to each number in the data array before taking logarithms, with the following result:

```
let z=log(1+insects)
condense z > zl
scat zl

    0.35|
        |
        |  *
        |
        |
        |                  *
        |      *
        |
        |                 *
        |
        |                   *
        |
        |         *
    0.10|--------------------------
       0.20                     1.40
```

We may conclude from this display that the logarithmic transformation has resulted in the relation between spreads and centers being reversed. In fact, the reciprocal transformation is even more powerful than the logarithm in reducing positive skewness in the distribution of a batch of numbers, as illustrated by the following example.

Imagine a single batch of numbers in which the lower quartile, median, and upper quartile are 1, 2, and 100, respectively - the distribution is positively skewed. Let us now see what happens to the median and quartiles when various transformations of the data are taken. We denote the transformed values as Q1, M, and Q2, respectively. These are still the median and quartiles of the reexpressed batch, by definition. As a measure of skewness of the distribution of the numbers in the batch, the quantity S = (M-Q1)/(Q2-Q1) is used, a value close to 1/2 indicating a symmetric distribution. The values of Q1, M, Q2, and S for various transformations are as follows:

	Q1	M	Q2	S
Raw data	1	2	100	0.010
Square roots	1	1.41	10	0.046
Cube roots	1	1.26	4.64	0.071
Fourth roots	1	1.19	3.16	0.088
Logarithms	0	0.30	2.00	0.150
Recip. sq. roots	1	0.71	0.10	0.326
Reciprocals	1	0.5	0.01	0.495

Notice that as we move from square roots to logarithms to reciprocals, S continually increases. We can thus order the transformations in terms of their ability to reduce positive skewness. In this hierarchy reciprocal powers are more extreme than ordinary powers, and logarithms fall between the two.

For the insect counts we may conclude from the last two plots that the most effective transformation for evening out the spreads is somewhere between the square root and the logarithm. Possibly cube roots are indicated. Going back to the schematic comparison display, we obtain the following schematic plot of the cube roots of the insect counts:

```
let z=insects↑Ø.333
compare z
```

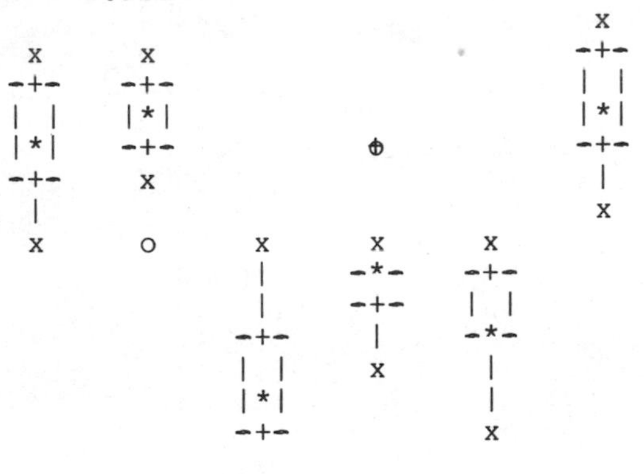

One conclusion which emerges from this analysis is that the batch contains two numbers that are relatively small. Looking at the data, we see that they are both zeros. The stem-and-leaf plot of the square roots also revealed them. It may be worth investigating to see if the two agricultural units involved were special in some way.

EXERCISES

1. The following data arose from an experiment to determine nitrogen-binding capabilities for three groups of laboratory mice. The animals were treated for 28 days with bovine serum albumen (BSA) antigen, and on the 29th day the amounts of BSA nitrogen bound in micrograms per milliliter of undiluted mouse serum in each animal were measured. (Source - article by R.E Dolkart, B. Halperin, and J.Perlman, "Comparison of antibody responses in normal and alloxan diabetic mice," Diabetes, Vol 20, 1971, pages 162-167.) Compare the responses in the three groups. Can any conclusions be made?

Normal	Alloxan diabetic	Alloxan diabetic treated with insulin
155.76	390.72	82.50
282.00	46.20	99.66
197.34	468.60	97.66
297.00	86.46	150.48
115.50	174.02	242.88
126.72	132.66	67.98
119.46	13.20	227.70
29.04	498.96	130.68
252.78	167.64	73.26
122.10	62.04	17.82
349.14	127.38	19.80
109.90	275.88	100.32
143.22	176.22	71.94
64.02	145.86	133.32
25.54	108.24	464.64
85.80	275.88	36.96
122.10	50.16	46.20
454.85	72.60	34.32
655.38		43.56
13.88		

2. The data listed arose from an experiment in learning. The responses are the numbers of syllables correctly anticipated by individuals after 34 minutes of practice. The subjects are grouped according to the length of rest interval between trials and the total number of trials. (Source - a book by Quinn McNemar, Psychological Statistics, Wiley, 1949, page 243.) Compare the responses in the five groups. State the conclusions reached.

Group	1	2	3	4	5
Rest Interval (min)	8	3.5	2	1.25	0
Number of trials	5	8	11	14	29
	5	8	9	11	17
	5	7	3	12	16
	1	4	9	15	18
	5	4	10	11	11
	8	7	5	10	15
	1	7	11	8	9
	2	5	9	13	18
	2	6	6	13	13
	2	8	7	5	12
	8	14	6	7	15
	4	8	16	11	8
	1	5	12	12	13
	3	1	11	12	7
	4	5	15	9	15
	4	8	13	16	15
	2	5	4	7	13

3. The following data array gives the mean ages in days of Prussian infants who died within a year of birth in the designated years and groups. (Source - article by Raymond Pearl, "On the mean duration of life in individuals dying within a year after birth," _Biometrika_, Vol 4, 1906, pages 510-516.) Make schematic comparison plots of the four distributions and comment on the appearance of each. Make a table containing the five-number summaries. What conclusions can be made from these data?

	Male		Female	
	Legitimate	Illegitimate	Legitimate	Illegitimate
1882	111.71	96.39	117.89	104.17
1883	112.91	100.04	117.72	103.80
1884	114.07	98.86	119.70	103.09
1885	113.91	98.15	119.34	106.26
1886	117.15	101.79	121.46	108.91
1889	112.97	98.00	118.28	104.72
1890	114.62	100.53	122.99	109.05
1891	110.96	96.37	116.26	102.54
1894	113.61	98.15	119.23	104.16
1895	111.89	98.58	116.89	104.07
1896	111.67	95.85	116.83	101.79

4. The following data are the live birthrates for women in six age groups in the United States, for various years. (Source - Statistical Abstract of the United States, Bureau of the Census, 1975.) Compare the birthrates for women in the different age groups, using, if desirable, some transformation to even the spreads. What conclusions can be made from these data?

Age	15-19	20-24	25-29	30-34	45-39	40-44
1940	54.1	135.6	122.8	83.4	46.3	15.6
1945	51.1	138.9	132.2	100.2	56.9	16.6
1950	81.6	196.6	166.1	103.7	52.9	15.1
1955	90.5	242.0	190.5	116.2	58.7	16.1
1960	89.1	258.1	197.4	112.7	56.2	15.5
1965	70.4	196.8	162.5	95.0	46.4	12.8
1967	67.9	174.0	142.6	79.3	38.5	10.6
1968	66.1	167.4	140.3	74.9	35.6	9.6
1969	66.1	166.0	143.0	74.1	33.4	8.8
1970	68.3	167.8	145.1	73.3	31.7	8.1
1971	64.7	150.6	134.8	67.6	28.7	7.1
1972	62.0	131.0	118.7	60.2	24.8	6.2
1973	59.7	120.7	113.6	56.1	22.0	5.4

5. The data which follow are the medians of the illiteracy rates among states in various regions of the United States at selected censuses from 1900 to 1970. (Source - Statistical Abstract of the United States, Bureau of the Census, 1975, page 120.) Compare the percentages of illiterates in various years. What conclusions can be drawn from this analysis?

	1900	1920	1930	1950	1960	1970
New England	6.5	4.9	3.0	2.0	1.4	0.7
Middle Atlantic	6.5	5.7	4.1	2.9	2.2	1.1
East No.Central	4.8	3.2	2.2	1.9	1.5	0.8
West No.Central	4.6	1.9	1.4	1.5	0.9	0.6
South Atlantic	23.4	10.2	7.7	3.9	2.7	1.4
East So.Central	28.0	14.6	11.0	5.4	3.4	1.9
West So.Central	18.4	9.6	7.4	5.2	3.8	2.0
Mountain	5.8	3.0	2.5	2.0	1.0	0.6
Pacific	5.3	3.6	2.8	2.2	1.8	1.1

6. For the data listed in Exercise 2 of Chapter 1 (aircraft disaster fatalities), form 12 batches by separating the data by month of disaster. Compare the distributions. What other comparisons could be made with these data?

7. Collect some data that interests you, and analyze them according to the techniques proposed in Chapter 2.

PROGRAM COMPONENTS

APL functions and FORTRAN subroutines for numerical summaries ("condense") and schematic plots ("compare") are listed in this section. As stated in the text, "condense" depends on two parameters, "num," which can be set to 1, 2, 5, or 7 and which determines the numbers of statistics in the summary, and "miss" ("zm" in FORTRAN), which is the code for missing values. The default value for "num" used in the text is 2, in which case the summary consists of the median and the midspread only. The missing value code should default to some number which is unlikely to occur naturally in the data, such as the largest number the computer can process.

The program "compare" has 4 parameters associated with it - "depth" ("idp"), "miss" ("zm"), "ngap" and "atom". The parameter "depth" controls the vertical height of the display - a default of 20 lines is reasonable for most applications. The parameter "ngap" determines the number of spaces between the vertical schematic plots - a default of 3 gives a visually attractive display, but if a large number of batches need to be compared, "ngap" should be reduced so that the display will fit on the page. As in the case of "boxplot," the subprogram "fill" is used to fragment the algorithm.

APL FUNCTIONS

```
       ∇CONDENSE[⎕]∇
       ∇ Z←CONDENSE X;I;N;NR;U;Q
[1]    X←(2ρ(ρX),1)ρX
[2]    Z←''
[3]    I←0×NR←(ρX)[2]
[4]    N←ρU←U[⍋U←(U≠MISS)/U←X[;I+1]]
[5]    Q←0.5×U[⌊0.125×7+N×⍳7]+U[⌈0.125×1+N×⍳7]
[6]    →7+2×⌊NUM÷2
[7]    Z←Z,Q[4]
[8]    →4+11×NR=I←I+1
[9]    Z←Z,Q[4],Q[6]-Q[2]
[10]   →4+12×NR=I←I+1
[11]   Z←Z,U[1],Q[2],Q[4],Q[6],U[N],N
[12]   →4+14×NR=I←I+1
[13]   Z←Z,U[1],Q[ 1 2 4 6 7 ],U[N],N
[14]   →4+16×NR=I←I+1
[15]   →0×ρρ(⎕← '  MEDIANS:' )
[16]   Z←(NR,2)ρZ
[17]   →0×ρρ(⎕← 'MEDIAN     MIDSPREAD' )
[18]   Z←(NR,6)ρZ
[19]   →0×ρρ(⎕← ' MIN  LOQ  MEDIAN  UPQ   MAX  SIZE' )
[20]   Z←(NR,8)ρZ
[21]   ' MIN  LO8  LOQ  MEDIAN  UPQ  UP8  MAX  SIZE'
       ∇
```

```
       ∇COMPARE[⎕]∇
       ∇ Z←COMPARE X;U;L;Y;M;C;J;K;P
[1]    Y←(MISS≠,X)/,X
[2]    X←(÷ATOM+U-⎕←L)×((⎕←U←⌈/Y)-DEPTH×L←⌊/Y)+X×DEPTH-1
[3]    M←(÷ATOM+U-L)×(U-DEPTH×L)+MISS×DEPTH-1
[4]    Z←(DEPTH,(NGAP+3)×C←(ρX)[2])ρ' '
[5]    J←0
[6]    K←(J←J+1)×NGAP+3
[7]    P←DEPTH FILL P[⍋P←(P≠M)/P←X[;J]]
[8]    Z[;¯2+K]← ' |¯⊥_⊤|I                          ' [1+P]
[9]    Z[;¯1+K]← ' -⊤=⊥= =|× ○○○○○○○○○ ⊛⊛⊛⊛⊛⊛⊛⊛⊛' [1+P]
[10]   Z[;K]← ' |¯⊥_⊤|I     23456789  23456789' [1+P]
[11]   →6×⍳J<C
[12]   Z←⊖Z
       ∇
```

FORTRAN SUBROUTINES

In both of the following subroutines, the data are stored in the two-dimensional array, x, whose dimensions are nr (rows) and nc (columns). In "compare," an additional one-dimensional array of size n (=nr*nc) is needed for temporary storage.

```
      subroutine condense(x,nr,nc,num,zm)
      dimension x(nr,nc),z(999),q(7),h(9)
      data h/'med','sprd','min','lo8','loq','upq',
     +'up8','max','size'/
      do 7 j=1,nc
      nz=Ø
      do 1 i=1,nr
      if(x(i,j).eq.zm) goto 1
      nz=nz+1
      z(nz)=x(i,j)
1     continue
      call sort(z,nz)
      do 2 k=1,7
2     q(k)=(z(int(Ø.99+k*nz/8))+z(int(1.Ø+k*nz/8)))/2
      goto (3,4,5,6),int((num+2.)/2.)
3     write(6,8) q(4)
      goto 7
4     write(6,8) q(4),q(6)-q(2)
      goto 7
5     write(6,8) z(1),q(2),q(4),q(6),z(nz),nz
      goto 7
6     write(6,9) z(1),q(1),q(2),q(4),q(6),q(7),z(nz),nz
7     continue
8     format(5f9.3,i8)
9     format(7f9.3,i8)
      goto (1Ø,11,12),minØ(3,maxØ(1,int(num/2.)))
1Ø    write(6,13) (h(i),i=1,num)
      return
11    write(6,13) h(3),h(4),h(1),h(6),h(8),h(9)
      return
12    write(6,13) (h(i),i=3,5),h(1),(h(i),i=6,9)
13    format(8(3x,a4,3x))
      return
      end
```

```
      subroutine compare(x,y,n,nr,nc,idp,zm,ngap,atom)
      dimension x(nr,nc),y(n),iy(99),a(30),id(999)
      logical*1 frmat(11)
      data a/'   ','|*|','-+-','-*-','-+-','-*-','| |',
     +'=*=',' | ',' x ','   ',' o ',' o2',' o3',' o4',
     +' o5',' o6',' o7',' o8',' o9','   ',' @ ',' @2',
     +' @3',' @4',' @5',' @6',' @7',' @8',' @9'/
      ny=0
      do 1 i=1,nr
      do 1 j=1,nc
      if(x(i,j).eq.zm) goto 1
      ny=ny+1
      y(ny)=x(i,j)
1     continue
      call sort(y,ny)
      write(6,7) y(ny)
      xmin=y(1)
      encode(11,2,frmat) ngap
2     format('(30(',i1,'x,a3))')
      aa=(idp-1.)/(atom+y(ny)-y(1))
      b=1.-aa*y(1)
      do 5 j=1,nc
      nz=0
      do 3 i=1,nr
      if(x(i,j).eq.zm) goto 3
      nz=nz+1
      y(nz)=aa*x(i,j)+b
3     continue
      call sort(y,nz)
      call fill(y,nz,iy,idp)
      do 4 k=1,idp
4     id(j+nc*(k-1))=iy(idp+1-k)
5     continue
      do 6 k=1,idp
6     write(6,frmat) (a(1+id(j+nc*(k-1))),j=1,nc)
      write(6,7) xmin
7     format(f12.3)
      return
      end
```

CHAPTER 3

RELATIONS

I seem to have been only like a boy playing on the seashore, and diverting myself in now and then finding a smoother pebble or a prettier shell than ordinary, whilst the great ocean of truth lay all undiscovered before me.

Sir Isaac Newton, from
Brewster's Memoirs of
Newton, Vol. 2, Chap. 27

FITTING A STRAIGHT LINE

In Chapter 2 we saw how to compare different batches of data. This is a fruitful path, and there is much to be gained by pursuing it further, as Chapter 5 shows. However there is another important area that we need to know about to be effective at data analysis. Accordingly let us change direction.

So far we have dealt only with batches of single numbers. Very often, however, we are called on to analyze batches of pairs of numbers. In this case we are interested in discovering relations between the component variables. A typical question might be, "What is the relation between the level of unemployment and the inflation rate in Great Britain in the 20th century?" Another question might be, "What is the relation between weight and height of women aged 30-39 years in the United States?"

A natural way to begin to answer questions like these is to make a scatter plot of the data. In the preceding chapter we used a scatter plot to show the relation between spreads and centers of batches of data, with the aim of finding some transformation to remove this relation. When dealing with data that are intrinsically bivariate, that is, that involve a natural pair of variables, we are primarily interested in the relation itself, not just in its removal. The simplest relation between two variables occurs when the points in the scatter plot tend to congregate around a straight line. In this case the relation may be specified by the two parameters that define the equation of the line: the slope, which is the amount y increases when x increases by 1, and the y-intercept, which is the distance, measured upwards, from the origin to the point where the line cuts the y-axis.

To illustrate the concept of a straight-line relation between two variables, consider the following data array, taken from page 216 of a book by Carl A. Bennett and Norman L. Franklin (Statistical Analysis in Chemistry and the Chemical Industry, Wiley, 1954). These data arise from a chemical experiment to prepare a standard curve for the determination of formaldehyde by the addition of chromatropic acid and concentrated sulfuric acid and the reading of the consequent purple color on a spectrophotometer. The first number in each pair is the amount of carbohydrate used in milliliters, whereas the second number is the optical density.

```
0.1       0.086
0.3       0.269
0.5       0.446
0.6       0.538
0.7       0.626
0.9       0.782
```

Labeling the above data array "acid," we may obtain a scatter plot using the program "scat" (introduced in Chapter 2):

```
  scat acid

      0.80|                     *
          |
          |
          |                *
          |             *
          |           *
          |
          |
          |      *
          |
          |
          | *
      0.00|--------------------------
        0.00                      1.00
```

It is not necessary to look very hard at the plot to notice a tendency for the six points to congregate around a straight line. If we were able to fit a straight line to the points, we would have a concise formulation of the relation between the optical density and the amount of carbohydrate used. Since it only takes two points to determine a line uniquely, the question arises of how to best fit the line. One way of fitting a straight line uniquely to a set of (x,y) points may be described as follows:

(a) Divide the points into three nonoverlapping regions, according to their x-values, in such a way that the three regions contain equal or nearly equal numbers of points. Compute the median of the x-values and the median of the y-values in each of the outer regions. Call these values (xB,yB) and (xT,yT), respectively. [In this example, each region contains two points, and the medial points are (0.2, 0.178) and (0.8, 0.704).]

(b) Compute the slope of the line joining the

points (xB,yB) and (xT,yT). By definition this is

(yT-yB)/(xT-xB)

(For our example, we get a slope of 0.877.)

(c) Compute the median of the differences y - slope*x, and take this as the y-intercept of the fitted line. (In this example we get 0.0065.)

This procedure is a modification suggested by John Tukey in lectures at Princeton University (see his Exploratory Data Analysis) of a method proposed by K.R. Nair and M.P. Shrivastava ("On a simple method of curve fitting," Sankya, Vol. 6, 1942-1944, pages 121-132). When fitting lines by hand, it is useful to

- mark the medial points (xB,yB), (xT,yT) on the scatter plot

- place a transparent straightedge next to these points

- move the straightedge parallel to itself until half the points are on each side, and then draw the line.

It is clear that this construction produces a straight line whose slope and y-intercept agree with the definition.

Suppose we have a program, "line," which routinely computes the slope and y-intercept for a line fitted to a set of (x,y) points using this procedure. Applying it to the data array "acid," we command

```
line acid
```

and get the response

```
slope:     0.8775  y-intercept:     0.0065
```

When scatter plots are made, it makes a difference which variable we choose as the y and which as the x. The program "scat" always assumes that the first column is x and the second column is y, in which case the second column is said to be plotted against the first. To avoid confusion we call the y variable, which is plotted vertically, the response, and the x variable, plotted horizontally, the carrier. The carrier is the variable we think of as being chosen, at least to some extent, by the experimenter, while the response is the value obtained from the experiment. It is not always obvious, however, which variable is the response and which is the carrier. There are circumstances where both variables are best thought of as responses; in these cases it is probably best to make two scatter plots, in which each response has a turn at being the y variable.

RESIDUALS

Once we have fitted a straight line to a set of (x,y) points, we have an equation of the form

y = y-intercept + slope*x

which summarizes the relation between the response y and the carrier x. Each time we put a different value of x from our data array into this equation, we will get a different y. The next question that arises is, "How well does the straight line fit the data?" To see how well the fitted line fits, there is a technique that is perhaps even more important to data analysis than reexpression, namely subtraction. In this case it involves subtracting the y-values of the fitted line from the y-values of the data, yielding the residuals from the fitted line. If these residuals are then plotted in a scatter plot against the fitted y-values, the goodness of fit of the fitted line shows up very clearly; this is similar to examining the original scatter plot under a strong magnifying glass.

To display residuals from a fitted line, let us introduce a parameter, "resids," into the program "line," having default Ø, but which causes the program to produce residuals when it is set to 1. We can now obtain a scatter plot of residuals by applying the program "scat" to the resultant array. For the data array "acid," we command

```
line acid {resids=1} > z
```

and obtain

```
slope:          Ø.8775Ø   y-intercept:        Ø.ØØ65Ø
```

If we then command

scat z

we obtain

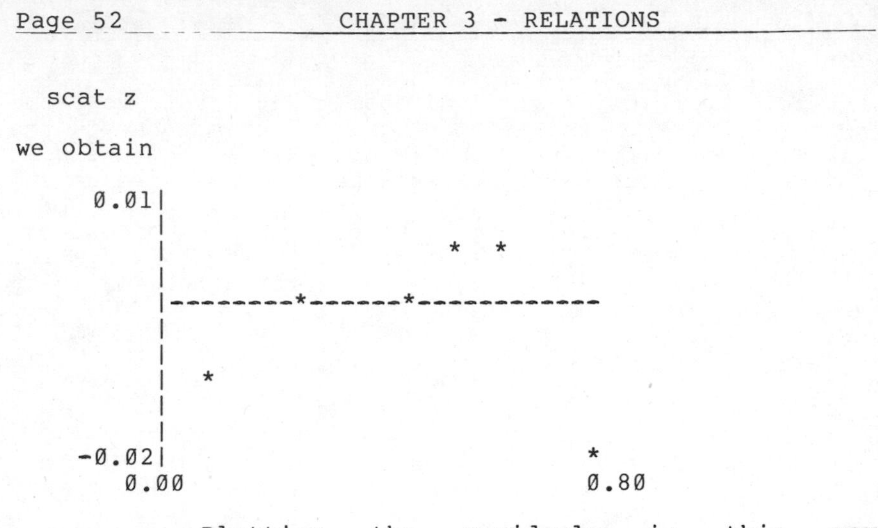

Plotting the residuals in this way highlights the fact that, although our fitted line describes most of the data well, there is one value, the one corresponding to the largest carbohydrate amount, that is out of line with the others. It is possible that this is a bad value, caused by the failure of the recording equipment, or by an error made by the experimenter in writing down the result. On the other hand, since the value corresponds to the extreme carbohydrate amount, there may be a real departure from the straight line relation at this end of the scale. In any case the plot of the residuals raises questions that may have interesting theoretical implications, and it is worth pursuing the matter further to discover the reason for the apparent discrepancy.

As another example consider the following data on stopping distances in feet (y) for various initial speeds in miles per hour (x), for 50 motorists. These are taken from Table 10 of a book by M. Ezekiel (Methods of Correlation Analysis, Wiley, 1930).

(4,2), (4,10), (7,4), (7,22), (8,16), (9,10), (10,18),
(10,26), (10,34), (11,17), (11,28), (12,14), (12,20),
(12,24), (12,28), (13,26), (13,34), (13,34), (13,46),
(14,26), (14,36), (14,60), (14,80), (15,20), (15,26),
(15,54), (16,32), (16,40), (17,32), (17,40), (17,50),
(18,42), (18,56), (18,76), (18,84), (19,36), (19,46),
(19,68), (20,32), (20,48), (20,52), (20,56), (20,64),
(22,66), (23,54), (24,70), (24,92), (24,93), (24,120),
(25,85)

Suppose we have these data stored in our computer with the label "cars." The scatter plot is obtained using the command

```
  scat cars
```

which yields

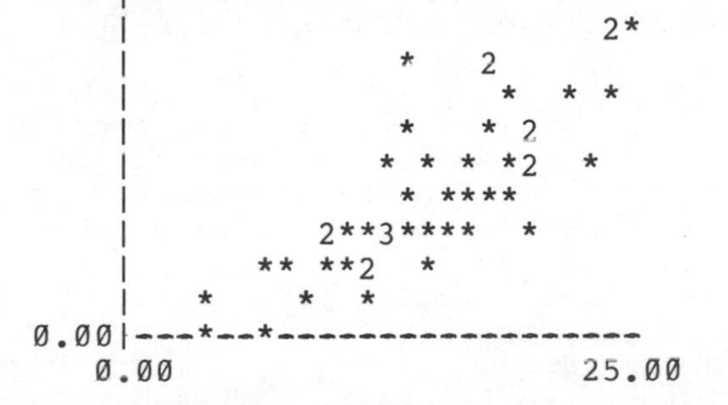

(When two or more points coincide, the number of incidences is used in place of the usual symbol.) After we fit a straight line, the plot of residuals against fitted values is obtained by the commands "line" and "scat," as follows:

```
  line cars {resids=1} > z

slope:        4.32000   y-intercept:       -24.84000

  scat z {width=40;depth=20}
```

```
    60.00      |
               |
               |          *
               |                      *
               |               *
               |          *
             * |               *
               | *   *  *  *    *      2
             * |  *  **
         ------|-*---*-22-*---**-*-------*-----
               |   *  *** * * *  2  *  *
               |       *   ** ****
               |           *    *    *
               |                 *
               |
   -40.00      |
  -20.00                                    100.00
```

(We assume that parameters "width" and "depth" can be used with "scat." Default values are 30 and 15, respectively. The preceding display is magnified to give greater resolution.)

Examination of this residual plot reveals some tendency, not evident in the original scatter plot, for the residuals to increase as the x-values (speeds) increase. This suggests that a straight line is probably not the best description of the distance-speed relation.

In some cases, particularly when the number of points being plotted is small, the line-fitting procedure described in the preceding section leaves residuals that still have some tilt - or tendency to track a straight line with detectable slope or other unbalanced curve - remaining in their scatter plot. In these cases it is desirable to repeat the line-fitting procedure on the plot of residuals versus x-values, and to recompute the slope as the sum of the original slope and the slope in this residual plot. What we are doing here is applying a concept, which we shall usefully apply repeatedly, of removing pattern from residuals and transfering it into the fitted relation. In future we assume that this iteration is done when fitting a straight line to (x,y) pairs.

TRANSFORMING THE RESPONSE

In Chapter 1 we showed how single batches can be transformed to obtain a more symmetric distribution, and thus simplify the estimation of the center. In Chapter 2 we saw how simultaneous transformation of several batches can remove a relation between spreads and centers of batches, facilitating comparison. We now come to a third situation requiring such reexpression. The application in this case is to a batch of pairs of numbers in which there is some relation between the components of each pair.

As a first example consider the following data of average weight (lb.) by height (in.) of American women aged 30-39. Height is the first member of each pair. (Source - The World Almanac and Book of Facts, 1975.)

(58,115), (59,117), (60,120), (61,123), (62,126),
(63,129), (64,132), (65,135), (66,139), (67,142),
(68,146), (69,150), (70,154), (71,159), (72,164)

The data are assumed to be stored under the lable "women." The scatter plot and residual plot after fitting a line are obtained, using the commands "scat" and "line," as follows:

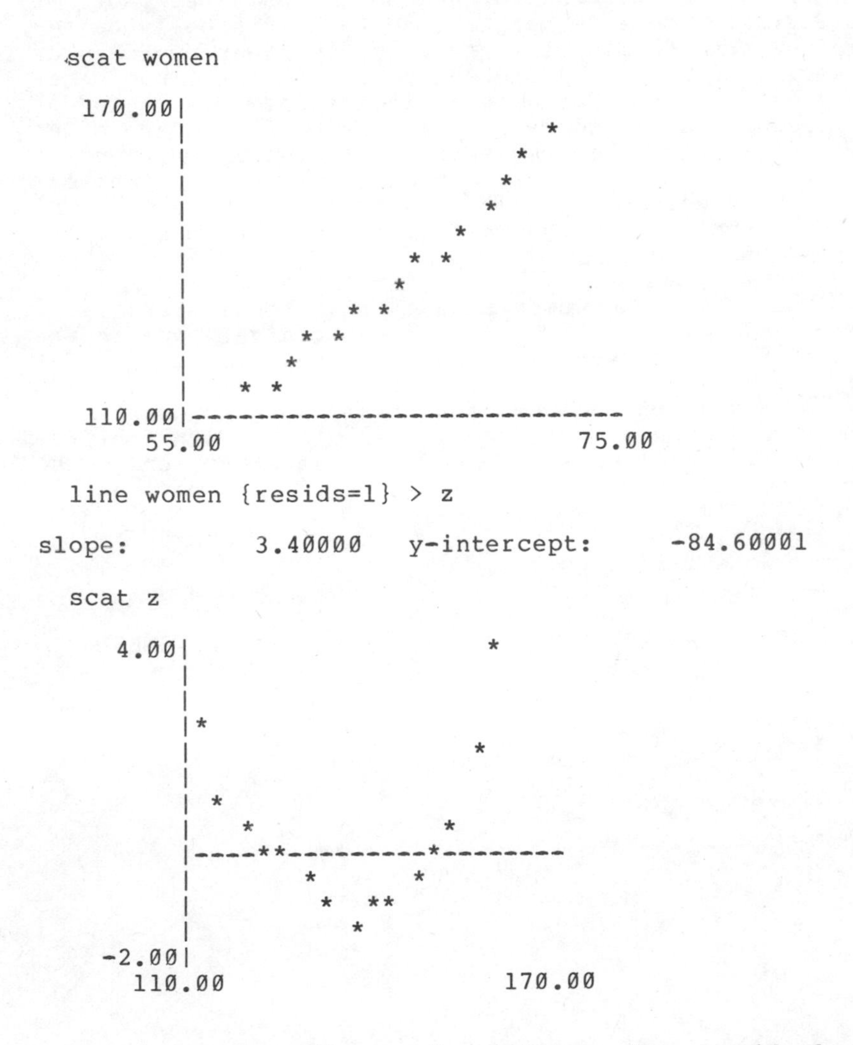

The effect of plotting the residuals is rather striking. The ordinary scatter plot of weights against heights appears to be reasonably straight, suggesting that as height increases weight increases in proportion. In contrast, the residual plot shows very clearly the departure from linearity in the relation between the two variables; in the lower range of heights, weight increases less rapidly with height than in the upper range of heights.

The presence of a bend in a scatter plot of residuals from a straight-line fit can usually be removed by transforming the y-values, in other words, by measuring them in a different scale. If the bend has a hollow in the middle, taking logarithms or roots of the responses is indicated. If the bend goes the other way, it can often be removed by taking squares, cubes, or exponentials, but it is usually preferable to reexpress the x-values by taking logarithms or roots. The choice of transformation will depend on what makes sense in the situation.

For the women's weights we find, after some trial and error, that fitting a straight line to the plot of the inverse square roots of the weights against the heights leaves unstructured residuals. (To transform just one column of a data array, assume we have another parameter, "col," associated with the program "let," which enables us to transform only the column specified by the value of "col.")

```
  let z=women↑-0.5 {col=2}
  line z {resids=1} > zl
```

```
slope:        -0.00107   y-intercept:        0.155555

  scat zl
```

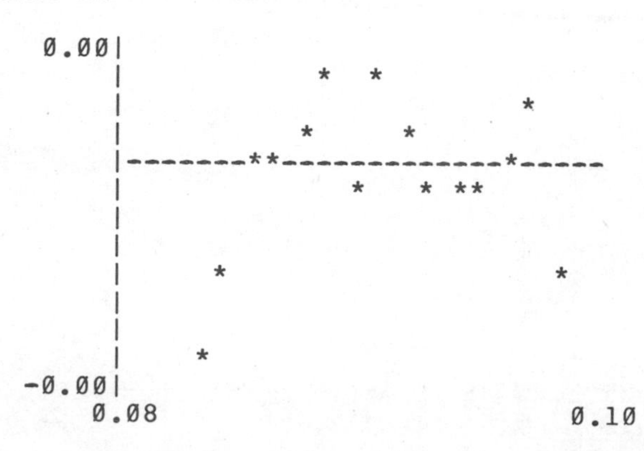

We may conclude that the relation between weight and height of U.S. women aged 30-39 years is, on the average, fitted reasonably well by the relation

1/squareroot(weight) = 0.156 - 0.001 * height,

or, squaring both sides of the equation and taking reciprocals, we obtain the formula

weight = 1/(0.156 - height/100)↑2

Why should such a relation hold? The physiologists may have some explanation, but we can offer none.

As a second example of straightening plots by transforming the response variable, we examine the relation between the vapor pressure of mercury in millimeters (of mercury) and temperature in degrees Celsius. These data, shown below and stored in our computer under the lable "vaporpressure," are obtained from page D-161 of the Handbook of Chemistry and Physics, (R.C. Weast, Editor, CRC Press, 1973).

Temperature	Pressure	Temperature	Pressure
0	0.0002	200	17.3
20	0.0012	220	32.1
40	0.006	240	57.
60	0.03	260	96.
80	0.09	280	157.
100	0.27	300	247.
120	0.75	320	376.
140	1.85	340	558
160	4.2	360	806.
180	8.8		

The scatter plot of vapor pressure against temperature is now obtained by invoking the command

```
scat vaporpressure
```

from which we obtain the display

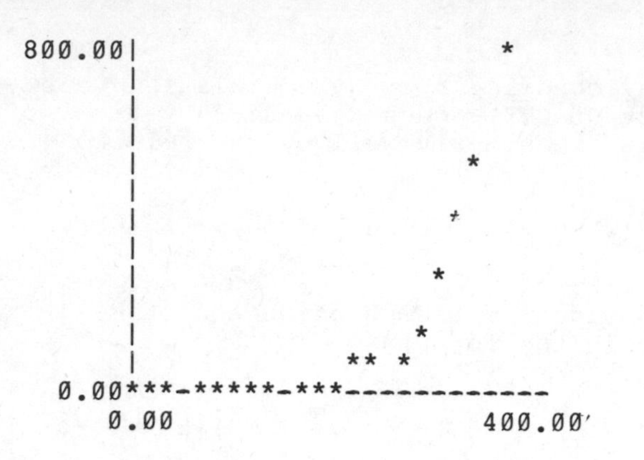

After some trial and error it is found that the transformation of vapor pressure that most effectively removes the bend from the scatter plot is the eighth root:

```
let z=vaporpressure↑0.125 {col=2}
line z {resids=1} > zl
```

slope: 0.00563 y-intercept: 0.30189

```
scat zl
```

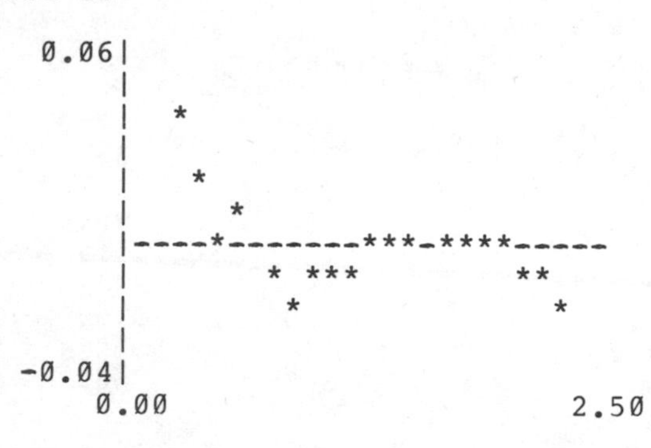

Although the bend in the scatter plot has been removed, considerable structure remains, and the result is hardly satisfactory. In the next section an improved fitting technique is demonstrated.

REEXPRESSING BOTH VARIABLES

In the preceding section we learned how to straighten a plot by transforming the response variable. In some situations it is necessary to transform both the response and the carrier, and in these situations, trying to straighten the plot by transforming only the carrier is like working with one arm tied behind one's back. As an illustration of the two-fisted technique, let us continue with the mercury vapor pressure example.

Data, in order to be transformed by powers or logarithms, should have a natural origin. It does not make sense to reexpress batches that include negative values, such as balances, or batches whose origin is arbitrary, such as dates. For the vapor pressure versus temperature situation, if we want to reexpress the temperatures, they should first be converted to numbers that are necessarily nonnegative. The lowest possible temperature is -273 degrees Centigrade, so it is reasonable to add 273 to the temperatures before transforming them. (On occasion, the number to add or subtract is not known from theoretical considerations, and must be estimated from the data.)

Using the command "let," we can add 273 to the first column of the data array "vaporpressure" as follows:

```
let z=vaporpressure+273 {col=1}
```

Recall that when the temperatures were not reexpressed, the best transformation of the vapor pressures to remove the bend from the plot of residuals was the eighth root. Since we now propose to transform temperature as well, let us start with some less extreme power transformation for y, and then find the power transformation for x that most effectively straightens the plot. We can then work our way down the power transformations for y until we find a pair of transformations that works well. After some trial and error, it is found that taking logarithms of y and reciprocals of x does fairly well. The analysis would proceed as follows:

```
let z1=log(z) {col=2}
let z2=z1↑-1 {col=1}
line z2 {resids=1} > z3
```

```
slope:    -3154.21265   y-intercept:       7.89973
```

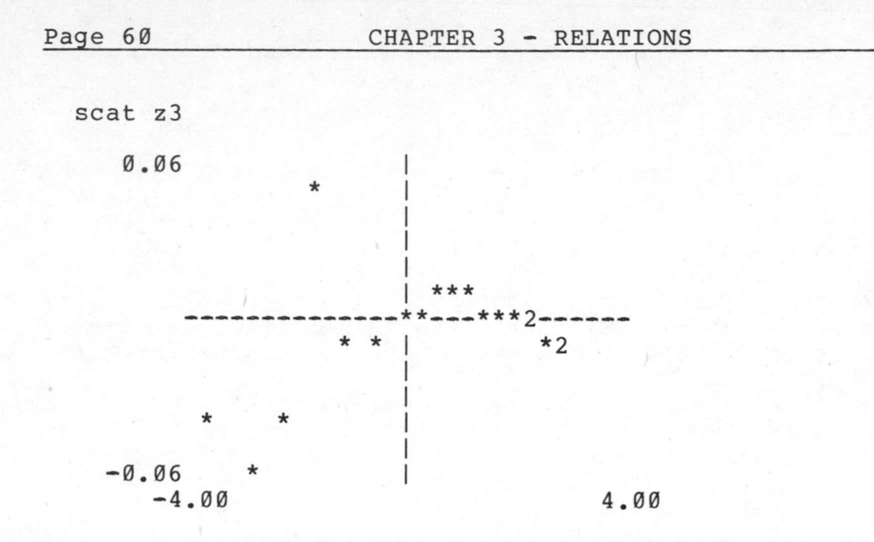

There is still some structure remaining in the residuals, but less than before, and the transformations of logarithms and reciprocals make more practical sense than eighth roots. This example is a good illustration of the use of data analysis to discover a physical law. As is well known in physics, the vapor pressure, P, of a liquid is given as a function of absolute temperature, T, approximately by the formula

$$\log(P) = a - b/T,$$

where a and b are constants that depend upon the properties of the liquid. In the case of mercury, the values a=7.90 and b=-3154 are obtained from the data analysis. (The logarithms are base 10 logarithms, which is our convention, unless otherwise stated.)

The pattern of the residuals in the preceding example is only one of the many different departures from randomness that residual plots can display. Recall the plot of residuals in the stopping distance versus speed relation. In that case there is a tendency for the residuals to increase in magnitude as the speeds increase. Just as there are standard ways of removing bends from scatter plots, there are standard ways of removing systematic trends in sizes of residuals. To illustrate this, let us write an equation for the relation that contains the residual term:

$$y(i) = a + b*x(i) + z(i).$$

Here the index i refers to the position of the pair (x,y) in the data array, a is the y-intercept and b the slope of the fitted line, and z(i) is the residual with index i. Now if there is a tendency for the residuals to increase in size as x increases, we can allow for this in one way by replacing the relation with

$$y(i) = a + b*x(i) + x(i)*Z(i),$$

where the Z(i) do not have a tendency to increase with x. This equation may be rewritten as

$$y(i)/x(i) = a/x(i) + b + Z(i),$$

or

$$Y(i) = A + B*X(i) + Z(i),$$

where A=b, B=a, Y(i) = y(i)/x(i), and X(i)=1/x(i). Consequently if we transform to the new variables (X,Y), we can fit a straight-line relation in which the residuals do not have a tendency to increase systematically over the range of fitted values.

After some experimentation with the data, we find that replacing the response by y/x and leaving the carrier x unchanged are effective in removing most of the structure from the residuals. Using the command "let" for the necessary manipulations, we obtain the following results:

```
  let y=cars {col=2}
  let x=cars {col=1}
  let z=x,(y/x)
  line z {resids=1} > zl

slope:          0.10000   y-intercept:          0.85000

  scat zl
```

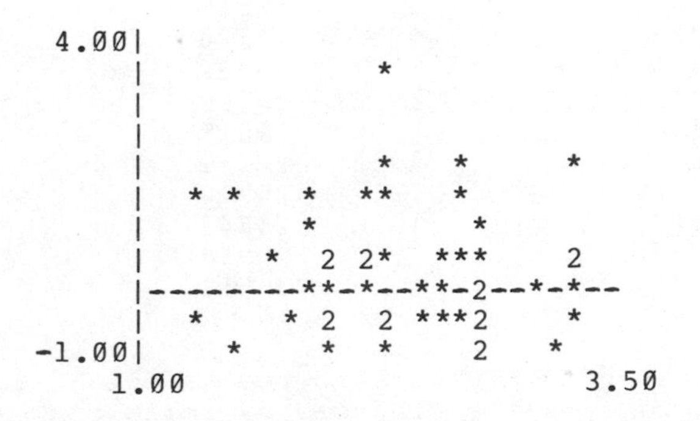

If we are satisfied that these residuals are random, we can now write the stopping distance versus speed relation as

distance = 0.85*speed + 0.1*(speed)↑2 + error,

where the error increases linearly with speed. In fact it is possible to construct a stochastic model that incorporates the above statistical relation, but this is more appropriately dealt with in a book on stochastic modeling.

TIME SERIES PLOTS

When the carrier is time, special methods are often useful for displaying and analyzing relations. In this case the data array is referred to as a *time series*. In time series analysis, it rarely if ever makes sense to transform the carrier. However it often does make sense to plot the response against its value at the previous time point, that is, its lagged value. Once this has been done, it is useful to look for a reexpression that straightens the resulting plot. As an illustration of this technique, let us examine the revenue passenger miles on U.S. passenger airlines for each year from 1937 to 1960. These numbers are obtained from the *F.A.A. Statistical Handbook of Aviation*, and are reproduced in a book by Robert G. Brown (*Smoothing, Forecasting and Prediction of Discrete Time Series*, Prentice-Hall, 1963) on page 427.

Year	Miles(000's)	Year	Miles(000's)
1937	412	1949	6753
1938	480	1950	8003
1939	683	1951	10566
1940	1052	1952	12528
1941	1385	1953	14760
1942	1418	1954	16769
1943	1634	1955	19819
1944	2178	1956	22362
1945	3362	1957	25340
1946	5948	1958	25343
1947	6109	1959	29269
1948	5981	1960	30514

Suppose we have these data stored with the label "airmiles". A scatter plot may be produced as follows:

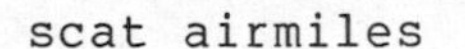

scat airmiles

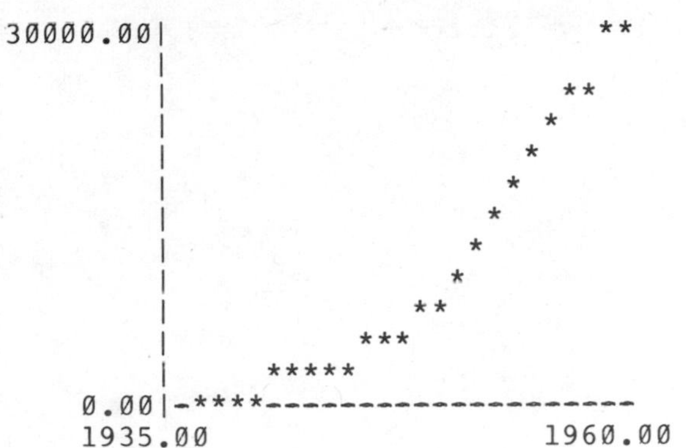

For data of this type the most common question of interest is how to predict the next value. To plot the response against its value at the previous time point, let us expand the definition of the program "let" to pick out a range of indices of an array, as follows:

```
let z=airmiles {col=2}
let y=z[2:24]
let x=z[1:23]
let z=x,y
```

We can now get a scatter plot of the response versus the lagged response as follows:

```
scat z
```

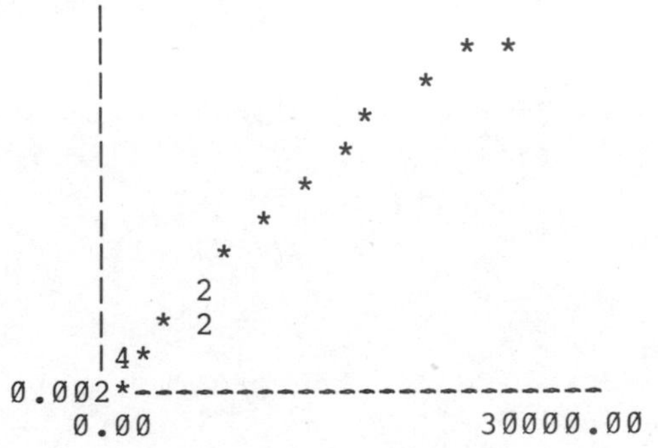

After we fit a straight line to these points, the residual plot has the following shape:

```
  line z {resids=1} > zl

slope:          1.12412   y-intercept:       202.42566

  scat zl
```

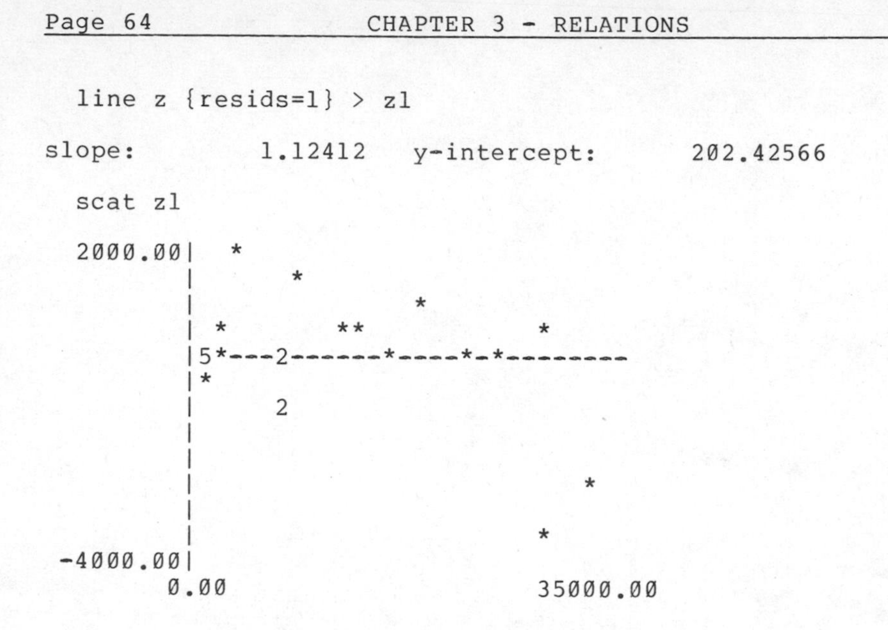

The residual plot tells us that the straight-line relation fits fairly well; however there are two points with large negative residuals, indicating that in two years the volume of air travel increased by less than would be expected from the linear relation. The linear relation predicts a yearly increase of 12.4% of the volume from the previous year, plus a fixed increase of 202,400 miles.

As a final example, consider the population of the United States at the 19 decennial censuses from 1790 to 1970. The populations in millions are as follows:

Year:	1790	1800	1810	1820	1830	1840	1850	1860	1870	1880
Pop.:	3.93	5.31	7.24	9.64	12.9	17.1	23.2	31.4	39.8	50.2

Year:	1890	1900	1910	1920	1930	1940	1950	1960	1970
Pop.:	62.9	76.0	92.0	105.7	122.8	131.7	151.3	179.3	203.2

Assume the 19 populations are stored, in order, in a data array labeled "uspop." Using the procedure outlined above, we obtain these results:

```
  let y=uspop[2:19]
  let x=uspop[1:18]
  let z=x,y
```

```
  line z {resids=1} > z1

slope:        1.09612   y-intercept:        5.67593

  scat z1
```

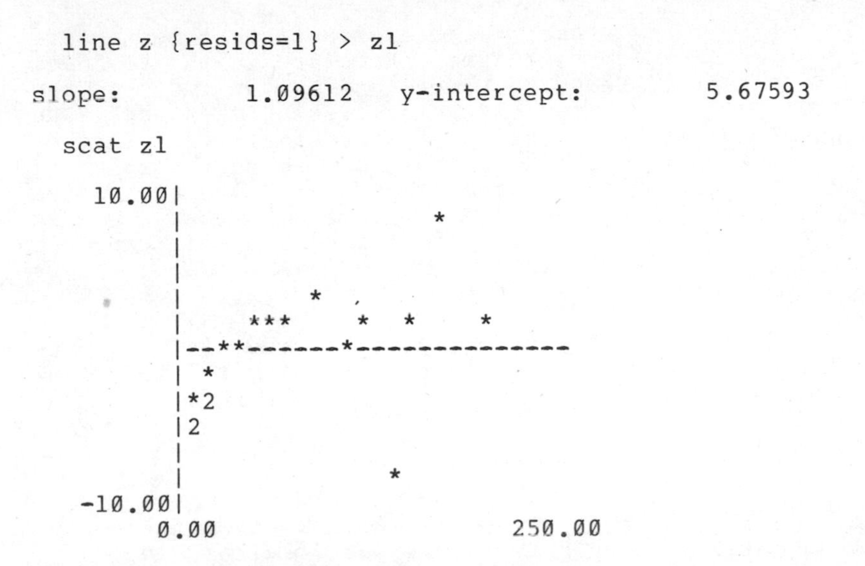

This plot of residuals shows quite a bit of structure. We try to remove or reduce this structure by transforming the population. It is found, after a little experimentation, that taking the reciprocal square root of the values and repeating the analysis on this transformed variable give a reasonably patternless residual plot:

```
  let z1=z↑-0.5
  line z1 {resids=1} > z2

slope:        0.83676   y-intercept:        0.00869

  scat z2
```

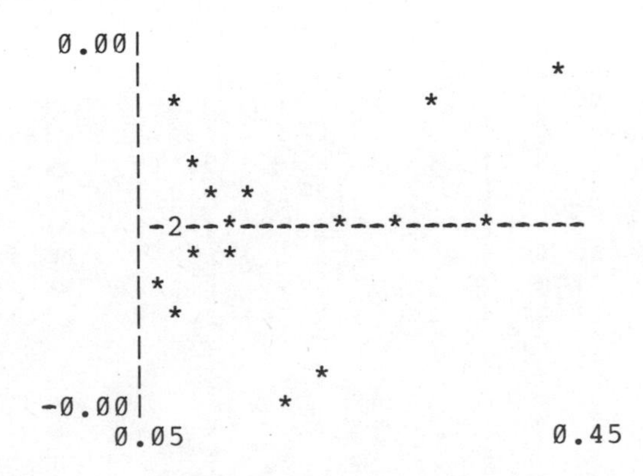

In this case the reexpression has been quite successful in reducing the structure in the residual plot, as well as reducing the relative magnitudes of the residuals. The analysis has also picked up some features not at first evident in the data. The largest two negative residuals correspond to the censuses of 1850 and 1860. The fact that these are in successive decades is suggestive. One wonders whether there is some reason for the low population increase in the period from 1840 to 1860. Historians or demographers may have a possible explanation.

The fact that the data led us to the reciprocal square root transformation yields the relation

$$y(i+1) = 0.0087 + 0.837*y(i),$$

where $y(i)$ is the reciprocal square root of the population size at census i. It is possible to construct a stochastic model for the U.S. population growth from this, which is similar to a certain model proposed by Raymond Pearl and L.J. Reed ("On the Rate of Growth of the Population of the UnIted States since 1790 and its Mathematical Representation," _Proceedings of the National Academy of Sciences_, Vol. 6, 1920, pages 275-288).

EXERCISES

1. The following data are the distances of the right upper pole calyx to the spine for 25 normal children whose ages are also given. (Source - an article by G. Friedland, R. Filly and B.W. Brown, Jr., in _Pediatric Radiology_, Vol. 2, 1974, pages 29-37.) Plot the distance against age for each child, and fit a straight line to the graph. By examining the residuals, determine the goodness of fit of the relation. Can you find an improved relation?

Age in Years	Distance in Millimeters
2	20
3	18
4	22 ,25
5	17, 20. 20, 22
6	21, 22
7	20, 20, 22, 24
8	18, 25, 33
9	27, 31
10	18, 24, 34
11	25, 28

2. Data on percent Democrat of major party vote for 24 Northeastern and Central States in four U.S. Presidential elections are given. (Source - The World Almanac and Book of Facts, 1975, pages 736-746.) Plot the 1972 values against those for each of the other three elections, obtaining three separate graphs. Which graph is best fitted by a straight-line relation? Discuss the implications of this analysis for election forecasting.

	1960	1964	1968	1972
Colorado	45.1	61.6	45.0	35.6
Connecticut	53.7	67.9	52.8	40.6
Delaware	50.8	61.1	47.8	39.7
Illinois	50.1	59.5	48.4	40.7
Indiana	44.8	56.2	43.0	33.5
Iowa	43.3	62.0	43.5	41.3
Kansas	39.3	54.6	38.8	30.3
Maine	43.0	68.8	56.2	38.6
Maryland	53.6	65.5	51.0	37.9
Massachusetts	60.4	76.5	65.7	54.5
Michigan	51.0	66.8	53.7	42.6
Minnesota	50.7	63.9	56.6	47.2
Nebraska	37.9	52.6	34.8	29.5
New Hampshire	46.6	63.9	45.8	35.2
New Jersey	50.4	66.0	48.8	37.4
New York	52.6	68.7	52.9	42.0
North Dakota	44.5	58.1	40.6	36.5
Ohio	46.7	62.9	48.7	39.0
Pennsylvania	51.2	65.2	51.9	39.8
Rhode Island	63.6	80.9	66.9	47.0
South Dakota	41.8	55.6	44.0	45.8
Vermont	41.4	66.3	45.2	36.8
West Virginia	52.7	67.9	54.8	36.4
Wisconsin	48.1	62.2	48.0	45.0

3. The following figures are the inflation and unemployment levels in the United States from 1959 to 1969. (Source - an article by Roger B. Spencer, "The Nation Plans to Curb Unemployment and Inflation," Federal Reserve Bank of St. Louis Review, April 1973, page 5.) Plot the inflation level against the unemployment rate for each year. By suitable use of transformations, fit a straight-line function to the data. Comment on the result of the analysis.

Year	Inflation	Unemployment
1959	0.7	5.4
1960	1.6	5.4
1961	1.0	6.6
1962	1.2	5.5
1963	1.2	5.5
1964	1.4	4.9
1965	1.7	4.4
1966	2.9	3.7
1967	2.9	3.7
1968	4.3	3.6
1969	5.4	3.5

4. Data on gross national product, and its three main components - personal consumption, gross private domestic investment, and government purchases - for the United States from 1929 to 1955, in constant (1947) billions of dollars, are tabulated. (Source - a book by Edmund Whittaker, Economic Analysis, Wiley, 1956, page 355.) Plot private investment against personal consumption, and fit a straight line to the resultant graph. What happens to the parameters in the fitted line if the war years (1942-1945) are omitted? Comment on the goodness of fit of the straight line in each case. What conclusions would you draw from the analysis?

Year	GNP	Pers.Cons.	GPDI	Govt.Purch.
1929	149.3	107.3	26.8	13.5
1930	135.2	100.9	17.9	15.2
1931	126.6	98.0	12.0	15.9
1932	107.6	88.9	3.3	15.1
1933	103.7	86.6	2.1	14.9
1934	113.4	91.5	4.3	17.3
1935	127.8	97.3	13.6	17.3
1936	142.5	107.6	15.2	20.3
1937	153.5	111.5	22.5	19.6
1938	145.9	109.8	12.1	22.1

1939	157.5	116.3	16.8	22.8
1940	171.6	122.5	22.8	24.0
1941	198.2	130.9	28.9	37.3
1942	223.6	128.1	14.7	81.8
1943	248.9	131.4	7.4	114.2
1944	268.2	135.9	9.2	127.1
1945	263.1	145.2	13.0	107.8
1946	233.8	162.4	32.4	33.9
1947	232.2	165.0	29.7	28.6
1948	243.9	168.0	38.8	34.8
1949	241.5	172.3	28.1	40.3
1950	264.7	182.8	45.3	37.8
1951	282.9	183.6	45.2	51.7
1952	293.3	189.2	39.1	63.4
1953	306.5	197.4	39.6	69.8
1954	300.5	200.7	36.7	61.7
1955	318.8	213.9	45.3	58.4

5. The Zurich monthly relative sunspot numbers from January 1958 to December 1972 are given. (Source - World Data Center A for Solar-Terrestrial Physics, NOAA, Boulder, Colorado.) Fit a statistical relation to these data by plotting the number for each month against the number for the immediately preceding month, and fitting a straight line to the resultant graph. To what extent does this relation account for the variability in the data? Does a transformation improve the goodness of the fit?

	Jan	Feb	Mar	Apr	May	Jun	Jul	Aug	Sep	Oct	Nov	Dec
1958	202	165	191	196	175	172	191	200	201	182	152	188
1959	217	143	186	163	172	169	150	200	145	111	124	125
1960	146	106	102	122	120	110	122	134	127	83	90	86
1961	58	46	53	61	51	77	70	56	64	38	33	40
1962	39	50	46	46	44	42	22	22	51	40	27	23
1963	20	24	17	29	43	36	20	33	39	35	23	15
1964	15	18	16	9	10	9	3	9	5	6	7	15
1965	18	14	12	7	24	16	12	9	17	20	16	17
1966	28	24	25	49	45	48	57	51	50	57	57	70
1967	111	94	112	70	86	67	92	107	77	88	94	126
1968	122	112	92	81	127	110	96	109	117	108	86	110
1969	104	120	136	107	120	106	97	98	91	96	94	98
1970	112	128	103	110	128	107	112	93	100	87	95	84
1971	91	79	61	72	58	50	81	61	50	52	63	82
1972	62	88	80	63	80	88	76	77	64	61	42	45

6. Using the data from Exercise 5 of Chapter 1, analyze

the relation between longevity and gestation period for the animals cited. Comment on the result of this analysis.

7. Collect some data comprising pairs of observations, and analyze them using the techniques outlined in Chapter 3.

PROGRAM COMPONENTS

This section contains listings of the programs "line" and "scat." In FORTRAN, "line" has a single parameter, "iresids," associated with it, which determines whether or not residuals are produced (1 for residuals, 0 for none). The program "scat" has five parameters - "wid" ("iwd" in FORTRAN), "dep" ("idp"), "ndivx," "ndivy," and "atom". The parameters "wid" and "dep" control the horizontal and vertical dimensions of the display (defaults used in the text are 30 and 15, respectively), while "ndivx" and "ndivy" are the desired number of units on the x- and y-axes, respectively (defaults used are 4,4). The scaling algorithm is such that the divisions on the axes occur at whole numbers, although these are not explicitly printed, except at the extremes of the plot.

APL FUNCTIONS

The function "LINE" accepts either a vector or a two-column matrix array as its argument. If the argument is a vector of size N, the carrier variable is constructed to be the first N positive integers.

The function "SCAT" accepts a vector or a matrix with up to nine columns as its argument. If the argument is a vector of size N, this is plotted against the first N positive integers. If a matrix argument is given, its second, third, and so forth, columns are plotted against its first column on the same axes.

```
      ∇SCAT[⎕]∇
      ∇ W←SCAT Z;N;X;Y;C;R;U;S;L;I;J;K;UT;CL;G;D;B;A;O;V
[1]   →3×ι(×/N)>+/N←ρZ
[2]   Z←⍉(2,ρZ)ρ(ιρZ),Z←,Z
[3]   Y←Z[;1+ιC←¯1+(ρZ)[2]]
[4]   R←ρZ←X←,Z[;1]
[5]   L←U←S←2ρ0
[6]   J←1+0×ρ(D←NDIVX,NDIVY),B←WID,DEP
[7]   UT←10*⌊10⍟CL←ATOM+((U[J]←⌈/Z)-S[J]←⌊/Z)÷D[J]
[8]   S[J]←UT×⌊S[J]÷UT←UT[1↑⍋|CL-UT←( 1 2 5 )×UT]
[9]   U[J]←UT×⌈U[J]÷UT
[10]  L[J]←1+G×⌊(B[J]-1)÷G←(U[J]-S[J])÷UT
[11]  Z←,Y
[12]  →7×ι3>J←J+1
[13]  A←(⌽L)ρ0
[14]  X←1+⌊0.5+(L[1]-1)×(X-S[1])÷U[1]-S[1]
[15]  Y←1+⌊0.5+(L[2]-1)×(Y-S[2])÷U[2]-S[2]
[16]  I←1
[17]  →20×ι1<C
[18]  A[Y[I;1];X[I]]←10⌊A[Y[I;1];X[I]]+1
[19]  →18+6×R<I←I+1
[20]  J←1
[21]  D←0=V←A[Y[I;J];X[I]]
[22]  A[Y[I;J];X[I]]←(10×V>K+1)+((K+1)×K=V)+(K←35-2×J)×D
[23]  →21×ιR≥I←I+1
[24]  →21×ιC≥J←J+I←1
[25]  O←(⌽ρA)⌊1⌈1+⌊0.5+(L-1)×S÷S-U
[26]  A[;O[1]]←A[;O[1]]+36×0=A[;O[1]]
[27]  A[O[2];]←A[O[2];]+35×0=A[O[2];]
[28]  W←' ×23456789⌹LL̲KK̲JJ̲II̲HH̲GG̲FF̲EE̲DD̲CC̲BB̲AA̲-|' [1+⊖A]
[29]  (⍕ 'RANGE OF X: ' ),⍕S[1],U[1]
[30]  (⍕ 'RANGE OF Y: ' ),⍕S[2],U[2]
      ∇
```

```
      ∇LINE[⎕]∇
      ∇ Z←LINE W;X;Y;N;M;P;I;J;K;X1;Y1;X2;Y2;F;SL;YI
[1]   F←(N←ρX←X[P←⍋X←W[;1]])ρK←SL←0
[2]   X1←0.5×X[I←⌊0.5×M+1]+X[J←⌈0.5×1+M←⌊0.5+N÷3]
[3]   X2←0.5×X[N+1-I]+X[N+1-J]
[4]   Y1←0.5×Z[I]+(Z←Z[⍋Z←M↑Y←W[P;2]-F])[J]
[5]   Y2←0.5×Z[I]+(Z←Z[⍋Z←(-M)↑Y])[J]
[6]   F←(SL←SL+(Y2-Y1)÷X2-X1)×X
[7]   K←K+1
[8]   →4×ιK<2
[9]   YI←0.5×Y[⌊0.5×N+1]+(Y←Y[⍋Y←W[P;2]-F])[⌈0.5×N+1]
[10]  Z←⍉(2,N)ρF,W[;2]-F←YI+SL×W[;1]
[11]  (⍕ 'SLOPE: ' ),(⍕SL),(⍕ ' Y-INTERCEPT: ' ),(⍕YI)
[12]  Z←((ρZ)×RESIDS≥1)ρZ
      ∇
```

FORTRAN SUBROUTINES

In the subroutine "scat" data are stored in x, of size nr, and y, a matrix array of size (nr,nyc), where nyc is the number of graphs to be plotted. An array z of size n=nr*nyc is needed for temporary storage. The subroutine, "scale", performs the scaling routine.

```
          subroutine scat(x,y,z,nr,n,np,m,nd,iwd,idp,
          +ndivx,ndivy,atom)
          dimension x(nr),y(nr,np),z(n),m(nd),im(29),f(17)
          logical*1 f
          data im/' ','*','2','3','4','5','6','7','8','9',
          +'$','h','p','g','o','f','n','e','m','d','l','c',
          +'k','b','j','a','i','-','|'/
          do 1 i=1,nd
1         m(i)=0
          do 2 i=1,nr
2         z(i)=x(i)
          call sort(z,nr)
          sx=z(1)
          ux=z(nr)
          call scale(sx,ux,iwd,ndivx,atom)
          do 3 i=1,nr
          do 3 j=1,np
3         z(j+np*(i-1))=y(i,j)
          call sort(z,n)
          sy=z(1)
          uy=z(n)
          call scale(sy,uy,idp,ndivy,atom)
          iox=max0(1,1+int(0.5+(iwd-1)*sx/(sx-ux)))
          ioy=max0(1,1+int(0.5+(idp-1)*sy/(sy-uy)))
          if(np.gt.1) goto 5
          do 4 i=1,nr
          in=1+int(0.5+(idp-1)*(y(i,1)-sy)/(uy-sy))
          ++idp*int(0.5+(iwd-1)*(x(i)-sx)/(ux-sx))
4         m(in)=min0(10,m(in)+1)
          goto 7
5         do 6 j=1,min0(8,np)
          k=9+2*(9-j)
          do 6 i=1,nr
          in=1+int(0.5+(idp-1)*(y(i,j)-sy)/(uy-sy))
          ++idp*int(0.5+(iwd-1)*(x(i)-sx)/(ux-sx))
          if(m(in).gt.k+1) m(in)=10
          if(m(in).eq.k) m(in)=k+1
          if(m(in).eq.0) m(in)=k
6         continue
7         do 8 j=1,idp
8         if(m(j+idp*(iox-1)).eq.0) m(j+idp*(iox-1))=28
```

```
      do 9 i=1,iwd
9     if(m(ioy+idp*(i-1)).eq.Ø) m(ioy+idp*(i-1))=27
      write(6,1Ø) uy,(im(1+m(idp*i)),i=1,iwd)
1Ø    format(f9.2,16Øa1)
      do 11 j=2,idp-1
11    write(6,12) (im(1+m(1-j+idp*i)),i=1,iwd)
12    format(9x,16Øa1)
      write(6,1Ø) sy,(im(1+m(1+idp*(i-1))),i=1,iwd)
      encode(17,13,f) iwd-12
13    format('(f12.2,',i2,'x,f11.2)')
      write(6,f) sx,ux
      return
      end
```

```
      subroutine scale(s,u,len,ndiv,atom)
      dimension unit(3),dif(3)
      cell=atom+(u-s)/float(ndiv)
      unit(1)=1Ø.**(int(99.+alog1Ø(cell))-99)
      unit(2)=2*unit(1)
      unit(3)=5*unit(1)
      do 1 i=1,3
1     dif(i)=abs(cell-unit(i))
      k=1
      if(dif(2).lt.dif(1)) k=2
      if(dif(3).lt.dif(2).and.dif(3).lt.dif(1)) k=3
      s=unit(k)*(int(99.+(s/unit(k)))-99)
      u=unit(k)*(99.-int(99.-(u/unit(k))))
      ndiv=(u-s)/unit(k)
      len=1+ndiv*((len-1)/ndiv)
      return
      end
```

In the subroutine "line," data are stored in two arrays, x and y, each of size n. Residuals are stored in an array z of length n, while a further array, w, also of size n, is needed for temporary storage.

```
      subroutine line(x,y,z,w,n,iresids)
      dimension x(n),y(n),z(n),w(n)
      do 1 i=1,n
      z(i)=x(i)
1     w(i)=y(i)
      call sort(z,n)
      xb=Ø.5*(z(int(Ø.99+n/6.))+z(int(1.Ø+n/6.)))
      x1=Ø.5*(z(int(Ø.99+n/3.))+z(int(1.Ø+n/3.)))
      x2=Ø.5*(z(int(Ø.99+2*n/3.))+z(int(1.Ø+2*n/3.)))
      xt=Ø.5*(z(int(Ø.99+5*n/6.))+z(int(1.Ø+5*n/6.)))
      slope=Ø
      j=Ø
2     j=j+1
      k=Ø
      do 3 i=1,n
      if(x(i).gt.x1) goto 3
      k=k+1
      z(k)=w(i)
3     continue
      call sort(z,k)
      yb=Ø.5*(z(int(Ø.99+n/6.))+z(int(1.Ø+n/6.)))
      k=Ø
      do 4 i=1,n
      if(x(i).lt.x2) goto 4
      k=k+1
      z(k)=w(i)
4     continue
      call sort(z,k)
      yt=Ø.5*(z(int(Ø.99+k-n/6.))+z(int(1.Ø+k-n/6.)))
      slope=slope+(yt-yb)/(xt-xb)
      do 5 i=1,n
5     z(i)=y(i)-slope*x(i)
      call sort(z,n)
      yint=Ø.5*(z(int(Ø.99+n/2.))+z(int(1.Ø+n/2.)))
      do 6 i=1,n
      z(i)=y(i)-yint-slope*x(i)
6     w(i)=z(i)
      if(j.lt.2) goto 2
      write(6,7) slope,yint
7     format('slope:',f15.5,'   y-intercept:',f15.5)
      if(iresids.eq.Ø) return
      write(6,8) (yint+slope*x(i),z(i),i=1,n)
8     format(2f15.5)
      return
      end
```

CHAPTER 4

ASSAYS

One must learn
By doing the thing; for though you
think you know it
You have no certainty, until you try.
Sophocles, Trachiniae

EXPERIMENTAL DESIGN

In Chapters 2 and 3 we began to chart two paths to obtaining insight into the structure of data; one involves the comparison of different batches, the other involves fitting relations between components of a multidimensional batch. Before continuing along these paths in Chapters 5 and 6, it seems appropriate to pause and focus on applications of the techniques already introduced. Fortunately, there is an important field of research in which the themes of Chapters 2 and 3 come together - biological assay.

On page 1 of the book Probit Analysis (Cambridge University Press, 1947), D.J. Finney defines biological assay, in its broadest sense, as "the measurement of the potency of any stimulus, physical, chemical or biological, physiological or psychological, by means of the reactions which it produces in living matter." Since this kind of measurement almost always involves conducting an experiment in which data are collected at some cost and often in which animals are destroyed, it is important to design the experiment in such a way that sufficient information is obtained from a minimal amount of data. Consequently the statistician needs to be concerned with the theory of effective experimental design. As an illustration, consider the following experiment, which is discussed on pages 188-189 of Finney's book.

The purpose of the experiment is to estimate the potency of various constituents of orchard sprays in repelling honeybees. Individual cells of dry comb were filled with measured amounts of lime sulfur emulsion in sucrose solution. Seven different concentrations of lime sulfur ranging from a concentration of 1/100 to 1/1,562,500 in successive factors of one-fifth were used, as well as a solution containing no lime sulfur. The responses for the different solutions were obtained by releasing 100 bees into the chamber for two hours, and then measuring the decrease in volume of the solutions in the various cells.

The first issue to be decided is the arrangement of the solutions in the cells. If 50 cells are enough to give the bees freedom of movement, then it is wasteful to put solutions in more cells. It is also important to have a balanced experiment, in which the number of cells having solution is the same at each level of concentration. A further consideration, that always arises when dealing with cells which are spatially located, is to make sure that there are no directional biases. This can be done by using a Latin square layout, which is a rectangular array in which each treatment occurs once in each row and once in each column. With the eight treatments (from largest to zero concentration) labeled A, B, C, D, E, F, G, and H, the following Latin square design was used in the experiment: the data, in milligrams consumed, are given below the treatment.

```
 D   C   F   H   E   A   B   G
57  84  87 130  43  12   8  80

 E   B   H   A   D   C   G   F
95   6  72   4  28  29  72 114

 B   H   A   E   G   F   C   D
 8 127   5 114  60  44  13  39

 H   D   E   C   A   G   F   B
69  36  39   9   5  77  57  14

 G   E   D   F   C   B   A   H
92  51  22  20  17   4   4  86

 F   A   C   G   B   D   H   E
90   2  16  24   7  27  81  55

 C   F   G   B   H   E   D   A
15  69  72  10  81  47  20   3

 A   G   B   D   F   H   E   C
 2  71   4  51  71  76  61  19
```

Turning now to the analysis of these data, we can consider them as eight batches, with each batch containing the responses at a given concentration, and the programs "condense" and "compare" may be used for comparing the responses at the various concentrations. Suppose we have the data stored in a rectangular array called "bees," with the columns containing the responses in decreasing order of concentration. This data array is obtained by rearranging the data in the preceding tableau, so that the first column contains the observations from treatment A (12, 4, 5, 5, 4, 2, 3, 2), the second column contains those from treatment B, and so forth. Using the command

```
compare bees {depth=25}
```

we obtain the schematic plot

130.000

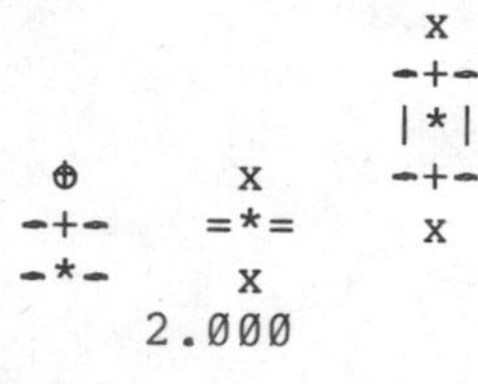

2.000

To get a numerical summary, we command

```
condense bees
```

and obtain

```
  4.000    2.500
  7.500    4.000
 16.500   10.000
 32.000   20.500
 53.000   33.000
 70.000   38.000
 72.000   13.000
 81.000   32.500
 med      sprd
```

Unlike the comparison situations we have previously encountered, this example has a quantitative variable - the concentration level - associated with the columns in the data array. Apart from the zero level, the levels are fixed multiples (of 5) of the lowest concentration. In biological assay it is typical to design the dosages of the stimulus in such a way that they are fixed multiples of a minimum dose. The reason for this convention becomes clear in the next section, where we deal with the problem of measuring potency.

The preceding example is discussed further in the following section. Before continuing this discussion, it is appropriate to mention another important application that arises in biological assay, namely, the problem of comparing the effectiveness of a new drug with a standard drug. Here one proceeds by treating subjects at various dosage levels of each drug. Since being dosed with the drug causes a change, possibly permanent, in the subject, it is not feasible to treat the same subject more than once. Thus the subjects, once chosen, should be allocated into two groups, and one group (the controls), treated with the standard drug, while the other group is treated with the new drug. These groups are further divided into subgroups, each of which is given a different dosage. There are a number of things the experimenter can control in designing such an experiment, in particular,

- the number and levels of the different dosages

- the number of subjects in each group

- the way subjects are allocated to the groups.

A problem can arise when there are systematic differences between subjects. For example, if the subjects are animals of both sexes, it may be unwise to put all the males in one group unless one is sure that there is no sex effect. With litters of laboratory animals such as mice, where siblings may have significant traits in common, it is usually best to assign each member of a given litter to a separate group, in which case the sizes of the litters constrain the number of treatment groups that can be compared. Factors such as these make experimental design an art. It is important for the data analyst to be aware of these considerations.

MEASURING POTENCY

As indicated in the preceding section, the procedure of bioassay typically involves dividing the subjects into two groups, and treating the control group with the standard and the other group with the new drug. The two groups are further divided into subgroups that are treated at different dose levels, the levels being simple multiples of a base dose. The data then consist of columns of responses for each drug, where each column corresponds to the responses at a given dose level. As an example of such a data array, we give the result of an assay of vitamin C activity of orange juice compared to ascorbic acid, taken from page 500 of a book by C.I. Bliss (The Statistics of Bioassay, Academic Press, 1952). The response is the average length of the odontoblasts (teeth) in each of 10 guinea pigs. Three dose levels - 0.5, 1, and 2 milligrams of juice and of acid in the juice - were used. The data array ("pigs") with schematic plots are now given.

Ascorbic Acid			Fresh Orange Juice			
4.2	16.1	23.6	15.2	19.7	25.5	
11.5	16.1	18.5	21.5	23.3	26.4	
7.3	15.2	33.9	17.6	23.6	22.4	
5.8	17.3	25.5	9.7	26.4	24.5	
6.4	22.1	26.4	14.5	20.0	24.8	
10.0	17.3	32.1	10.0	25.2	30.9	
11.2	13.6	26.7	8.2	25.8	26.4	
11.2	14.5	21.5	9.4	21.2	27.3	
5.2	18.8	23.3	16.1	14.5	29.4	
7.0	15.5	29.1	9.7	27.3	23.0	
-1	0	1	-1	0	1	log(dose)

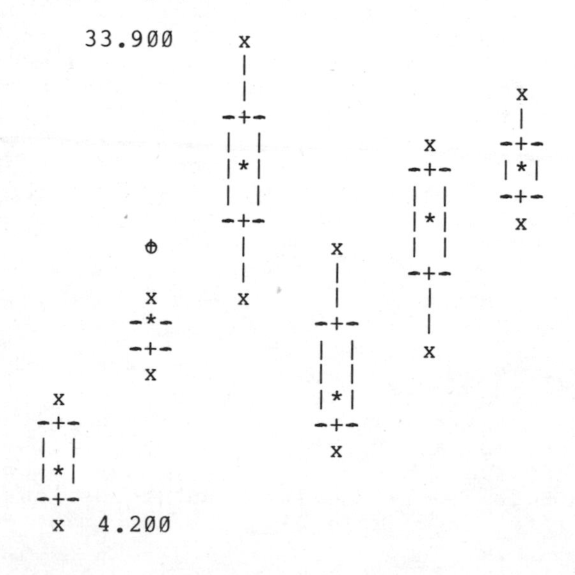

This display gives some idea of the response-dose relation for each substance. The responses for the orange juice, which is the unknown in this case, are generally higher than those for the ascorbic acid, so we may conclude that the potency of orange juice with respect to the standard is greater than one. The question is, "How much greater?"

To answer this question, we need to define precisely what we méan by potency. The definition is as follows:

> A new drug has potency P with respect to a standard drug if a dose D of the new drug behaves like a dose P*D of the standard.

When logarithms are used, this statement translates into saying that a log dose X of the new drug behaves like a log dose M+X of the standard, where M=log(P). [To see this, suppose X=log(D). Then from the definition, the corresponding log dose for the standard is log(P*D), which equals log(P)+log(D), which is M+X if M=log(P) and X=log(D).]

From this definition it follows that if the potency of an unknown drug with respect to a standard does not change with dose level, then the response-dose curve for the unknown drug (with the doses measured on a logarithmic scale) will be the same as that for the standard drug, except that it will be shifted by an amount M=log(P) to the right along the x-axis. So to measure the potency, we simply need to measure this shift, and to take its antilog.

To produce plots of the response-dose relations, we use the same procedure as in the preceding section; that is, we take the median of the responses in each group as a typical value for that group, and plot these values against the log doses. The medians are found by invoking the command "condense pigs," from which we obtain

```
  7.150    5.400
 16.100    2.100
 25.950    5.800
 12.250    6.400
 23.450    5.800
 25.950    2.800
 med       sprd
```

To plot the medians of the responses against the logarithm of the doses, we use the program "scat." To plot more than one response against a single dose, we use the program "scat" with a data array having more than two columns, where the first column is the column to be treated as a carrier and the other columns are treated as responses. Suppose we have such a data array, called "juice," with the coded logarithms of the doses in the first column, the acsorbic acid responses in the second column, and the orange juice responses in the third column. To plot the six points giving the two relations on a single graph, we command

```
scat juice
```

and get

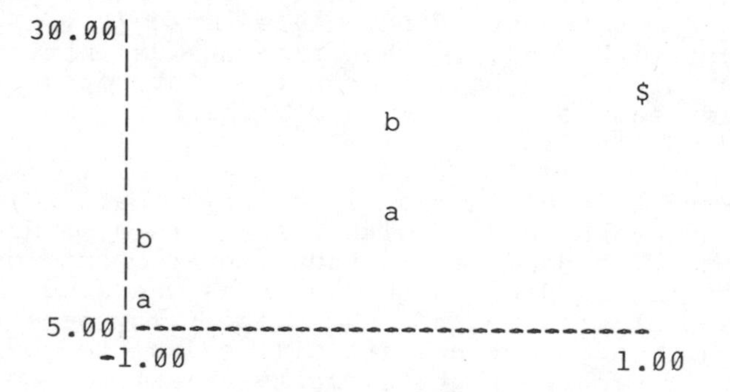

```
30.00|
     |
     |                                  $
     |                  b
     |
     |
     |                  a
     |b
     |
     |a
 5.00|-------------------------------
   -1.00                          1.00
```

The two points at the upper right, which correspond to the highest dose, are coincident, and are marked with the symbol $. Apart from the point corresponding to the orange juice response at the highest dose level, the response-dose relation is approximately linear. Ideally one wants to have a plot comprising a pair of parallel lines, because this makes it easy to measure the shift between the two curves that yields the potency. The experimenter has some control over this by choosing the dose levels and the responses appropriately. If one is satisfied that a pair of parallel straight lines can be fitted to the relation, the next question is that of choosing a fitting procedure. A simple way of doing this is described as follows:

(a) Fit a straight line separately to the plot of response versus log dose for the unknown and for the standard; let S be the average of these slopes.

(b) Calculate the residuals - response-S*log(dose) - for each relation; then compute the difference between the median of the residuals for the new drug and the median of the residuals for the standard; call this M. (It is not hard to see that M is a measure of the average distance between the fitted parallel lines, measured in the direction of the responses.)

(c) The estimate of the potency is antilog(M/S). [M is divided by S, the slope, to obtain the distance between the parallel lines measured in the horizontal direction. Since this axis is the log(dose) axis, it is necessary to take antilogarithms to bring it back to dose units.]

For the pig data the slope of the straight line fitted to the standard (ascorbic acid) is obtained, using the program "line," as 9.4. The commands that give this result are as follows:

```
let x=juice {col=1}
let y=juice {col=2}
let z=x,y
line z

slope:        9.40000  y-intercept:       16.55000
```

To calculate the slope for the unknown, we should really ignore the point corresponding to the response at log dose 1. The reason for this departure from linearity is probably a "ceiling" effect, which often occurs in bioassays, when increasing the dose beyond a certain level does not result in further change of the response. Since we only have two points left, we can calculate the slope simply as (y2-y1)/(x2-x1) = (23.45-12.25)/{0-(-1)} = 11.2. The average of the slopes is thus (9.4+11.2)/2 = 10.3. The residuals for the standard are accordingly

```
 7.15 - 10.3*(-1) = 17.45
16.10 - 10.3*0    = 16.10
25.95 - 10.3*1    = 15.65
```

with a median of 16.1, while those for the unknown are

```
12.25 - 10.3*(-1) = 22.55
23.45 - 10.3*0    = 23.45
```

having a median of 23.0. The log potency is thus {23.0-16.1}/10.3 = 0.67. In the experiment the doses differed by factors of 2, so we need the base 2 antilogs to get the potency, which is antilog{0.67*log(2)} = 1.59. The conclusion is that the orange juice is more than half again as potent as is its ascorbic acid.

FOLDED REEXPRESSIONS

In the honeybees experiment the question of interest is the potency of the lime sulfur in repelling the bees, as a function of the dosage level. We could have answered this question, to some extent, by using just eight cells in the brood comb, plotting the amount of solution consumed in each cell versus the concentration, and fitting a straight line to the resultant points, possibly after transforming the variables using the methods outlined in Chapter 3. The reason for having eight cells at each dosage level was to obtain more accurate estimates of the responses at the various levels, by averaging out the errors that can arise in the determinations. As we saw in Chapter 1, the median value is a good statistic to measure the central or typical value of a batch of numbers, provided the distribution of these numbers is reasonably symmetric. Therefore it is reasonable to compute the median of the responses at each dosage level, and to plot these median responses versus the concentrations to determine the potency. (Examination of the schematic plots does not reveal any consistent departure from symmetry within the batches, so we are justified in using the medians as typical values.)

Before we make a plot of this relation, it makes sense to reexpress the concentrations in logarithms, since these concentrations are multiples of each other, and taking logarithms will make the transformed values equispaced. We will replace the zero concentration by a very small positive number, 1/37,500,000, which is two factors of 5 smaller than the smallest concentration (this is to avoid taking logarithms of zero).

Taking logarithms to base 5, we create a data array, labeled "sulfur," containing the logarithms of the concentrations in the first column and the medians of the responses (given in the result of the command "condense bees") in the second column. The scatter plot has the following shape:

```
scat sulfur

 100.00                                 |
                                        |
                                        |
        *                               |
               *   *                    |
                                        |
                                        |
                      *                 |
                                        |
                                        |
                          *             |
                                        |
                                        |
                             *          |
                                 *   *
   0.00---------------------------------|
     -8.00                            0.00
```

Let us now consider the appropriate transformation of the response. So far in considering transformations of data we have kept within the family of power and logarithmic reexpressions. These transformations are reasonable when the data being reexpressed have a natural origin at zero and no upper limit. However, when the data have a natural origin at zero <u>and</u> an upper limit, it is often more informative to look at proportions rather than raw values.

In the present case the responses have a natural origin at zero (since it is not possible for the bees to consume a negative quantity of solution), but there is also an upper limit to the response, which occurs when the concentration of lime sulfur is zero. Thus it is natural to measure the response in terms of the fraction of solution consumed per unit amount of what would be consumed if the concentration were zero. This latter amount is not known with certainty, but it may be estimated from the data as 81 milligrams, the median response at zero concentration. Converting the responses to proportions by dividing by this value, we now have data with two natural origins, one at zero and the other at unity.

As suggested by John Tukey in lectures at Princeton University, one family of transformations appropriate for working with proportions is the family of folded roots and logarithms

$$y = x^p - (1-x)^p, \text{ if } p \neq 0,$$

$$y = \log\{x/(1-x)\}, \; p=0.$$

Recall that the rationale for taking roots and logarithms of nonnegative data is to stretch the data near the origin with respect to the rest of the range of the data. The reason for the folded reexpression of proportions is similar; we want to stretch the values near 0 and 1 with respect to those near 0.5. This is a reasonable thing to do if we believe that it is harder to shift a proportion from 0.01 to 0.02 or from 0.98 to 0.99 than it is to shift it from 0.49 to 0.50.

Let us take folded transformations of the proportional responses in the array "sulfur," in each case fitting a straight line to the resulting plot, and examining the residuals. The value of p that appears to give the least structured plot of residuals is zero. The analysis, which is carried out on only seven points (since we have already used the point corresponding to zero concentration to estimate the extreme response), is obtained as follows (using an obvious notation for the manipulation):

```
  let z=sulfur[1:7,*]
  let zl=log(z/(81-z)) {col=2}
  line zl {resids=1} > z2

slope:        -0.42927   y-intercept:        -1.43997

  scat z2

     0.20                |
        *                |
                         |     *
                         |
        ----*--*----|-*----------
                       * |
                         |
                         |
                         |
    -0.25                |        *
       -1.50                     1.50
```

The effect of changing the extreme response value is worth noting. If we choose a value of 76 instead of 81, the following results are obtained:

```
  let zl=log(z/(76-z)) {col=2}
  line zl {resids=1} > z2

slope:        -0.45752   y-intercept       -1.46753

  scat z2
```

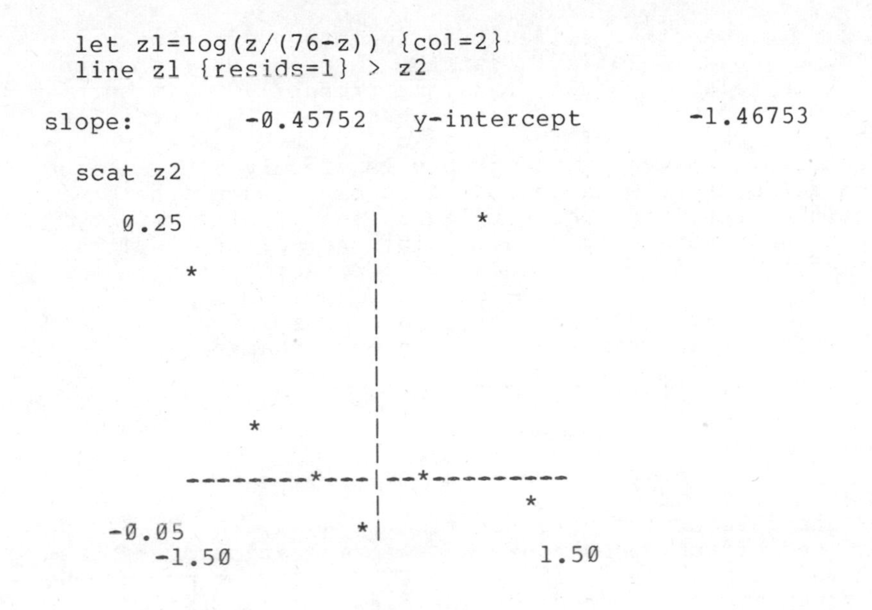

We have now quantified the relation between the response and the dose as

$$\log(\text{response}/(R-\text{response})) = a + b*\log(\text{dose})$$

where R is the maximum response or the response at zero dose level, and a and b are constants. The calculations were performed with logarithms to base 10 on the left-hand side and with logarithms to base 5 on the right-hand side; changing to any base does not alter the form of the relation. In the future when transforming according to folded logarithms, we shall assume that natural logarithms are used, in which case the folded reexpression is called a <u>logit</u>. If logits are used, the relation may be turned around to give

$$\text{response}/(R-\text{response}) = \exp\{a+b*\log(\text{dose})\}$$

that is,

$$\text{response} = R/\{1+A*\text{dose}^{B}\}$$

where A and B are constants related to a and b.

Another folded transformation often used in response-dose relations is the probit tansformation, which is related to the Gaussian probability integral, is not expressible in simple terms, and which requires tables. For most of the range of values usually encountered, however, the probit is closely approximated by a folded root with p=0.14. In many cases there is little difference between logits and probits, since it may be shown that logits arise in the limit, suitably scaled, of folded roots when p approaches zero.

We shall encounter more examples of the need for folded transformations in the next chapter.

EXERCISES

1. The data exhibited arose from an assay to determine the effectiveness of a new preparation of testosterone propionate on the growth of capons. The response for each of the 30 birds is the increase in (height + length) of comb after the bird is injected with a dose of the chemical, the responses being coded to take integer values. (Source - D.J. Finney's book, Statistical Method in Biological Assay, Hafner, 1964, page 139.) Calculate the potency of the test preparation with respect to the standard.

	Standard			Test		
Dose (micrograms)	20	40	80	20	40	80
Responses:	6	12	19	6	12	16
	6	11	14	6	11	18
	5	12	14	6	12	19
	6	10	15	7	12	16
	7	7	14	4	10	15

2. In a prolactin assay 24 pigeons were weighed before treatment and then dosed on six successive days by subcutaneous injection of either a standard or a test preparation. The birds were then sacrificed and their body weights before and crop-gland weights after treatment, in grams, are tabulated here. (Source - D.J. Finney's book, page 326.) Compute the potency of the test preparation, with respect to the standard, from these data. Carry out at least one analysis to estimate the potency.

Initial body weights						Final crop-gland weights					
Standard dose			Test dose			Standard dose			Test dose		
1.25	2.5	5.0	1.25	2.5	5.0	1.25	2.5	5.0	1.25	2.5	5.0
490	490	490	510	480	450	3.8	5.3	8.5	2.8	4.8	6.0
530	530	530	510	510	520	3.9	10.2	14.4	6.5	4.7	13.0
440	460	410	500	480	500	4.8	8.1	5.4	3.5	5.4	8.3
490	510	430	520	500	530	6.2	7.5	8.5	3.6	7.4	6.0

3. The response of rats to vitamin D can be measured by a procedure called the bone line test, which measures the antirachitic activity of the vitamin on a scale from 0 to 12 units. The responses for 60 rats treated at six dose levels are given. These data are the sums of the values from the line tests, as measured by two independent observers to avoid bias. (Source - Finney's book, page 71.) Plot the response-dose relation and attempt to fit a straight line to this relation after suitable transformations of the variables. If successful in this quest, write the relation mathematically in terms of response as a function of dose.

Dose(i.u.)	0.32	0.64	1.28	2.15	4.30	8.60
Response	1	2	4	8	14	20
	4	0	9	17	14	21
	1	2	4	6	13	16
	0	4	13	14	19	21
	0	0	3	17	17	15
	1	5	7	16	17	14
	0	3	4	8	20	20
	2	4	4		18	20
	2	2	10		17	15
	2	6	12			20
	0	1	11			20

4. The following percentages relate to high school seniors in the United States taking the standardized SAT tests. Each student is counted once only, the most recent score being used, and the percentages are given with respect to the total number of students taking each test. (Source - Educational Testing Service, Princeton, N.J.) By plotting and examining the data, compare the distributions between males and females, and between verbal and mathematical tests. How would you go about quantifying these relations? Can you think of theories that would explain the differences? Examine also the time trends.

	<300	<400	<500	<600	<700	<300	<400	<500	<600	<700
	Males, Verbal					Females, Verbal				
71-2	6.9	32.3	64.8	88.1	98.1	7.0	32.6	65.5	89.1	98.4
72-3	8.3	34.7	67.8	89.7	98.7	8.3	34.9	69.2	90.9	98.8
73-4	8.1	35.0	67.5	89.3	98.4	8.4	35.7	69.3	90.7	98.7
74-5	9.7	37.4	70.5	91.4	98.7	10.0	38.2	72.9	92.7	99.0
	Males, Mathematical					Females, Mathematical				
71-2	3.2	19.3	46.8	76.2	94.4	5.4	28.7	62.3	88.4	98.5
72-3	3.3	19.8	47.3	77.6	95.1	5.5	29.0	62.1	89.3	98.8
73-4	3.2	20.7	48.9	77.1	94.3	5.2	30.2	63.4	88.4	98.3
74-5	4.0	22.8	50.5	78.5	94.7	6.6	34.2	66.6	90.2	98.7

5. Fit a statistical relation to the U.S. population data cited in Chapter 3, Section 5, by a folded transformation of the data plotted against time. Determine the goodness of fit of the relation by examining residuals. Compare the relation with the one derived in the text.

6. Transform the birthrate data given in Exercise 5 of Chapter 2 to proportions of completed fertility by cumulating across rows and then dividing by the numbers in the last column. Plot the resulting fractions with age group as the carrier, and find a folded reexpression that straightens the plot. Fit a straight line to this plot and discuss the goodness of fit.

7. Collect some data that interests you and analyze them according to the techniques suggested in this chapter.

CHAPTER 5

TABLES

Errors, like straws, upon the surface flow,
He who would search for pearls must dive below.
John Dryden, All for Love
(1678), Prologue

TWO-FACTOR ARRAYS

In Chapter 2 we presented a method for analyzing rectangular data arrays for which each column represents a different factor or classification. In many situations we encounter arrays in which both the rows and the columns represent meaningful factors, and in these cases we are interested in the relation between the response represented in the body of the table and each of the two factors. Such data arrays are called _two-way tables_ or _contingency tables_.

Data, both experimental and nonexperimental, frequently are represented in the form of two-way tables. When tables are designed to convey information, it may be rather wasteful to give all the numbers in a batch - where multiple numbers arise from the same batch it may be sufficient to give a summary measure of each batch using quantities such as the mean or the median. Since it is desirable to convey as much information as possible with as few numbers as possible, it is economical to display the numbers in the form of a rectangular array in which both the rows and the columns have meaning. In the exercises and examples in the preceding chapters we have seen several instances of two-way tables.

In a designed experiment it also may be desirable to collect data according to a two-way or greater than two-way layout, so that the maximum amount of information will be contained in the data. The two-way layout can also be used in a designed experiment to account for some of the random variation, thus enhancing our ability to discover real effects. In the pig-orange juice bioassay discussed in the preceding chapter, we had a rectangular array in which the column factor was the dose level and the row factor was the litter identity. Since the object was to determine the relation between the response and the dosage, the litter factor was not of particular interest. However the litters may have been different, causing greater variation (spread) in the responses in the different columns than would have otherwise occurred. In this case it would be desirable to take the litter factor into account when assessing the potency.

When both factors are to be taken into account, we need a way of summarizing the data such that the relation between the response and each factor is highlighted. As an illustration consider the following simple data array, consisting of just four values of mean monthly temperatures in California (source - The World Almanac and Book of Facts, 1975, page 245):

Los Angeles, January:	57 degrees	Fahrenheit	
Los Angeles, July:	73	"	"
San Francisco, January:	48	"	"
San Francisco, July:	63	"	"

A little examination of these numbers reveals that July is 16 degrees hotter than January in Los Angeles, while July is 15 degrees hotter than January in San Francisco. Furthermore, Los Angeles is 9 degrees hotter than San Francisco in January, and 10 degrees hotter in July. If we were asked to make a hypothesis from these data, it would be that July is about 16 degrees hotter than January, while Los Angeles is about 10 degrees hotter than San Francisco. Of course we need data from more locations before we can be confident about our hypothesis, but at least we have something to work with.

The hypothesis may be stated symbolically as

response
 = typical value + location effect + seasonal effect

In our example it would be reasonable to choose 60 as a typical value, +5 (Los Angeles) and -5 (San Francisco) as location effects, and +8 (July) and -8 (January) as seasonal effects. This would give rise to the hypothetical values 57, 73, 47, and 63, compared with the actual values of 57, 73, 48 and 63. The only discrepancy is in the third value. The error could be spread evenly over the four values, by choosing different values for the typical value and location and seasonal effects. Taking 60.25 as the typical value and choosing location effects as ± 4.75 and seasonal effects as ± 7.75 would give rise to the fitted values 57.25, 72.75, 47.75, and 63.25. If these estimates are rounded off to integers, the fit is perfect.

MEDIAN POLISH

In a 2*2 table, an additive relation can be fitted by straightforward trial and error. In a table with more than two rows and columns, a more general procedure is needed. One such procedure, suggested by John Tukey in lectures at Princeton University (see his _Exploratory Data Analysis_), is _median polish_.

Consider the following two-way table of demographic data, in which the responses are death rates (per thousand) in Virginia in 1940 and the factors are (1) age group and (2) sex and urban/rural class. (Source - a paper by L. Moyneau, S.K. Gilliam, and L.C. Florant, "Differences in Virginia Death Rates by Color, Sex, Age, and Rural or Urban Residence," American Sociological Review, Vol. 12, 1947, pages 525-535.)

Ages	Rural Male	Rural Female	Urban Male	Urban Female
50-54	11.7	8.7	15.4	8.4
55-59	18.1	11.7	24.3	13.6
60-64	26.9	20.3	37.0	19.3
65-69	41.0	30.9	54.6	35.1
70-74	66.0	54.3	71.1	50.0

To median polish this table, we first compute the median of each row; we obtain the values 10.2, 15.8, 23.6, 38.0, and 60.2. We then attach each median to the corresponding row, and subtract them from the numbers in each row, obtaining

1.5	-1.5	5.2	-1.8	10.2
2.3	-4.1	8.5	-2.2	15.8
3.3	-3.3	13.4	-4.3	23.6
3.0	-7.1	16.6	-2.9	38.0
5.8	-5.9	10.9	-10.2	60.2

Next we compute the medians of the numbers in each of the resulting columns, obtaining the numbers 3.0, -4.1, 10.9, -2.9, and, for the attached column, 23.6. These medians are then attached to the end of each column, and subtracted from the remaining numbers in the respective columns. This process yields

-1.5	2.6	-5.7	1.1	-13.4
-0.7	0	-2.4	0.7	-7.8
0.3	0.8	2.5	-1.4	0
0	-3.0	5.7	0	14.4
2.8	-1.8	0	-7.3	36.6
3.0	-4.1	10.9	-2.9	23.6

This tableau displays estimates of the five age effects in the right-most column, estimates of the four sex and urban/rural class effects in the bottom row, and an estimate of the typical value for the death rates in the bottom right-hand corner. Thus if we hypothesize that the death rates for these people in Virginia in 1940 may be represented as an additive combination of effects due to age and sex/class, the discrepancies from this relation (fitted as shown) are the residuals in the body of the tableau.

As might be expected, the age effects account for most of the variation in the death rates - the effects indicate that there is a difference of 50 deaths per thousand between the 50-54 year olds and the 70-74 year olds. The differences due to sex and urban/rural residence are comparatively minor, although the urban males have a somewhat higher average than the others.

This procedure of "taking out" medians bears a resemblance to the line-fitting procedure introduced in Chapter 3. In that case an improved fit could be obtained, on occasion, by repeating the procedure on the residuals from the initial fit. The same is true here: in fact, iteration - or <u>polishing</u> - is almost always desirable when fitting relations to two-way tables. To do this, we first calculate the medians of the residuals in each row (ignoring the last column), subtract them from the rows, and add them to the last column. In the present example the row medians of residuals are -0.2, -0.4, 0.6, 0, -0.9, and, for the bottom row, 0. (These numbers are rounded to one digit after the decimal point.) Next the medians of the columns (ignoring the numbers in the bottom row) are calculated, and subtracted from the numbers in the columns and added to the bottom row. In the present case these medians are -0.3, 0.2, 0.9, 0, and, for the right-most column, 0.6. The result of this polish is the following tableau:

-1.0	2.6	-6.0	1.3	-14.2
0	0.2	-2.9	1.1	-8.8
0	0	1.0	-2.0	0
0.3	-3.2	4.8	0	13.8
4.0	-1.1	0	-6.4	35.1
2.7	-3.9	11.8	-2.9	24.2

The process may now be repeated, until further iteration does not result in changing the tableau. In the present case two additional polishes are necessary to reach this equilibrium, and the final tableau is

-1.2	2.4	-7.6	1.3	-14.0
-0.1	0.1	-4.0	1.2	-8.7
0	0	0	-1.8	0
0	-3.4	3.6	0	14.0
5.0	-0.1	0	-5.2	34.1
2.9	-3.7	12.9	-3.0	24.0

Our analysis indicates a typical death rate of 24.0, plus age effects of -14.0, -8.7, 0, 14.0, and 34.1, plus sex/class effects of 2.9 for rural males, -3.7 for rural females, 13.0 for urban males, and -2.9 for urban females.

In general, we shall call a relation of this type a <u>row-plus-column</u> fit, since the data may be decomposed as

Response
= Typical Value + Row Effect + Column Effect + Residual,

or, in algebraic terms, as

$$y(i,j) = t + r(i) + c(j) + z(i,j),$$

where the indices (i,j) refer to the row and column, respectively. Thus y(i,j) is the response in the i-th row and j-th column, t is the typical value, r(i) is the i-th row effect, c(j) is the j-th column effect, and z(i,j) is the residual in the i-th row and j-th column.

SIZE OF RESIDUALS

The median polish procedure is designed to replace the original table by an associated residual table for which the medians of each row and each column are zero. Polishing can be thought of as "sweeping out" medians, and terminating only when there is nothing left to sweep out. Since the median of a batch of numbers is that number which, when subtracted from all the numbers in the batch, minimizes the sum of the absolute values, or magnitudes, of the residuals, it should be clear that each step in median polishing a two-way table reduces the sum of the magnitudes of the values in the table. Therefore it is natural to use this sum as a measure of how well the row-plus-column relation fits the data.

In the death rates example considered in the preceding section, the sum of the magnitudes of the residuals is 40.3 after the first sweep, 37.9 after the next iteration, and 37.0 when convergence is reached. The average size of the residuals is thus 37/20 = 1.85.

To measure the goodness of fit of the row-plus-column relation, we need something with which we can compare the size of residuals. One such benchmark is the spread of the distribution of the data, before the first sweep. Suppose we have the death rate data array stored under the label "deaths." Then we can get the midspread of the data by invoking the program "condense". Since the data array is rectangular, we need first to convert it to a one-dimensional array; otherwise "condense" will produce the midspreads of the columns of the batch, rather than the midspread of the whole batch. Assume that we can do this "unravelling" using the program "let" and a comma indicating the unravel operation. Thus we command

```
let z=,deaths
condense z
```

and obtain

```
25.600   31.000
med      sprd
```

The midspread of the entire batch is thus 31.0.

We can now compare the average size of the residuals with the midspread of the data from the original table. To be consistent, we should really be comparing the midspread of the distribution of the residuals with the midspread of the data, or, alternatively, the average size of the residuals with the average deviation of the original data from the center of their distribution. The midspread of the distribution of the residuals for the death rate data is calculated from the last tableau as {upper quartile - lower quartile} = 0.5 - (-1.9) = 2.6, so fitting the row-plus-column relation has reduced the midspread by a factor (31.0-2.6)/31.0, or 92%. On the other hand, subtracting the median, which is 25.6, from the numbers in the original table, and summing the magnitudes of these residuals yields a total of 315.4, so the reduction in residual size as indicated by this measure is (315.4-37.0)/315.4, or 88%.

Thus one can say that about 90% of the spread or variation in the data is accounted for by the row-plus-column relation. The relative size of residuals is not the only, or even the most important measure of efficacy of fit of the relation to the data. The absence of structure in residuals is equally important. In the next sections we talk about techniques for more meaningful displays of residuals from relations fitted to two-way tables.

The median polish procedure can be rather boring to perform by hand, and it is better if a computer is available to eliminate the drudgery. Suppose we have a program, called "medpolish," which routinely fits a row-plus-column relation to a two-way table, using median polish. Applying it to the array "deaths", we command

```
medpolish deaths
```

and get

```
    40.300
    38.250
    37.575
    37.238
   315.400
row effects:
   -13.963    -8.700     0.000    14.038    34.275
column effects:
     2.888    -3.712    12.813    -2.950
typical value:         24.012
```

The values produced by the program "medpolish" are the sums of magnitudes of residuals after each polish. These values differ slightly from the ones we obtained by hand calculation, because computer calculations are more accurate than hand calculations. The value 37.0 was not reached because the program "medpolish" ceases polishing when the change in the sum of the magnitudes of the residuals is less than 1%. The last value produced before the row and column effects is the sum of the magnitudes of the data after their median has been subtracted from each. If we want the computer to display the residuals, we simply change the parameter, "resids," to the value one. In the present case we would command

```
  medpolish deaths {resids=1}
```

and get

```
    40.300
    38.250
    37.575
    37.238
   315.400
row effects:
   -13.963     -8.700      0.000     14.038     34.275
column effects:
     2.888     -3.712     12.813     -2.950
typical value:         24.012

    -1.238      2.362     -7.462      1.300
    -0.100      0.100     -3.825      1.238
     0.000      0.000      0.175     -1.762
     0.063     -3.437      3.738      0.000
     4.825     -0.275      0.000     -5.337
```

When the "resids=1" option is used, the residuals will be displayed after the row effects, column effects, and typical value.

REEXPRESSING TWO-WAY TABLES

Our experience from the preceding examples suggests that we should be on the lookout for making an improvement in the analysis by transforming the data. When dealing with single batches, the object was to obtain a fairly symmetric distribution. For comparisons of several batches, evenness of spreads was the goal. For relations in (x,y) pairs, we sought reexpressions that would straighten the plot. In the case of a two-way table the transformation sought is one for which the row-plus-column relation fits the transformed data well. The "acid test" of how well a relation fits data is the absence of structure in the residuals. As an illustration consider the following two-way table, which gives selected personal consumption expenditures, in billions of dollars, for the United States in various years (source - The World Almanac and Book of Facts, 1962, page 756).

	1940	1945	1950	1955	1960
Food and Tobacco	22.2	44.5	59.6	73.2	86.8
Household Operation	10.5	15.5	29.0	36.5	46.2
Medical and Health	3.53	5.76	9.71	14.0	21.1
Personal Care	1.04	1.98	2.45	3.40	5.40
Private Education	.341	.974	1.80	2.60	3.64

Suppose we have these data stored in a computer with the label "consumption." Using the program "medpolish," and asking for residuals, we command

```
medpolish consumption {resids=1}
```

and obtain

```
   139.295
   139.295
   426.185
row effects:
    49.890    19.290      0.000     -7.260     -7.910
column effects:
    -6.180    -3.950      0.000      4.290     11.390
typical value:          9.710
   -31.220   -11.150      0.000      9.310     15.810
   -12.320    -9.550      0.000      3.210      5.810
     0.000     0.000      0.000      0.000      0.000
     4.770     3.480      0.000     -3.340     -8.440
     4.721     3.124      0.000     -3.490     -9.550
```

Examining these residuals, we do not need much time to spot a pattern in them - the residuals are positive in the top right-hand and bottom left-hand corners, and negative in the top left-hand and bottom right-hand corners. In contrast the residuals from the median polish of the death rate array were fairly randomly distributed in sign over the array. To spot this kind of structure when it is less obvious, we could replace the residuals by a code which indicates a property of the residuals; for example, we could code them by their sign. The coded residual displays for the death rates and for the consumption data, after fitting the row-plus-column relation in each case, are

```
Death Rates          Consumption

-  +  -  +           -  -  .  +  +
-  +  -  +           -  -  .  +  +
.  .  +  -           .  .  .  .  .
+  -  +  .           +  +  .  -  -
+  -  .  -           +  +  .  -  -
```

where + indicates a positive residual, . a zero residual, and - indicates one which is negative.

As an alternative we could code the residuals according to their size. One way of doing this is based on the quartiles; residuals below the lower quartile of their batch are assigned one code, those between the quartiles are assigned another, and those residuals above the upper quartile are given a third code. To calculate these quartiles, we can first store the residuals under some temporary label and then use the program "condense" to give a five-number summary. The commands and responses would be as follows:

```
medpolish deaths {resids=1} > z
let zl=,z
condense zl {num=5}

 -7.462    -1.500    0.000    0.706    4.825     20
 min       loq       med      upq      max       size

medpolish consumption {resids=1} > z
let zl=,z
condense zl {num=5}

-31.220    -3.490    0.000    3.210   15.810     25
 min       loq       med      upq      max       size
```

Assigning the symbols - , . , and + to those residuals in each data array below the lower quartile, between the quartiles, and above the upper quartile, respectively, we obtain the coded residual displays

```
Death Rates        Consumption
.  +  -  +         -  -  .  +  +
.  .  -  +         -  -  .  .  +
.  .  .  -         .  .  .  .  .
.  -  +  .         +  +  .  .  -
+  .  .  -         +  +  .  -  -
```

The advantage of using the latter display is that it highlights the larger residuals, and these contain almost all of any structure which remains in the data after the median polish.

The kind of pattern present in the residual map for the array "consumption" can be removed by reexpressing the data. To show this, we take logarithms of the expenditures, and then perform the two-way analysis by commanding

```
let z=log(consumption)
medpolish z {resids=1}
```

and obtaining

```
   1.130
   1.032
   1.032
  13.921
row effects:
   0.788      0.433      0.000     -0.561     -0.732
column effects:
  -0.425     -0.227      0.000      0.142      0.306
typical value:         0.987

   0.000      0.100      0.000     -0.053     -0.143
   0.026     -0.003      0.042      0.000     -0.061
  -0.014      0.000      0.000      0.017      0.031
   0.015      0.097     -0.037     -0.037      0.000
  -0.298     -0.040      0.000      0.017      0.000
```

To get the coded residual map using the second coding scheme, we calculate the quartiles of the residuals with the commands

```
medpolish z {resids=1} > z1
let z2=,z1
condense z2 {num=5}

 -0.298   -0.037    0.000    0.017    0.100      25
 min      loq       med      upq      max       size
```

and then we construct the residual display. We obtain

```
   .   .   .   -   -
   +   .   +   .   -
   .   .   .   .   +
   .   +   .   .   .
   -   -   .   .   .
```

The effect of taking logarithms has been to remove most of the structure from the residuals. There is a bonus: the proportion of variation accounted for by the row-plus-column relation, in terms of sum of magnitudes of residuals, has risen from (426.4-139.1)/426.4 = 0.67 to (13.9-1.0)/13.9 = 0.93. We may conclude that fitting the row-plus-column relation to the logarithms of the data is more satisfactory than fitting it to the raw data.

This analysis seems rather <u>ad hoc</u> in the

sense that we provided no motivation for the logarithmic transformation. Why use logarithms instead of square roots or reciprocals? In fact, as in the preceding situations where gains can be made through reexpression, there is a simple general procedure for finding an appropriate reexpression when fitting row-plus-column relations to two-way tables. Suggested by John Tukey in lectures at Princeton University (see his Exploratory Data Analysis), it involves examining the scatter plot of the residuals versus the comparison values, which are defined as the products (row effect)*(column effect)/(typical value). If the points in this scatter plot show a tendency to congregate around a straight line whose slope is one, a logarithmic transformation is indicated. To see why this is so, suppose that the residuals exactly conform to this linear relation. Then, using algebraic notation, it follows that

$$z(i,j) = r(i)*c(j)/t$$

where z(i,j) is the residual in row i and column j of the table, r(i) is the i-th row effect, c(j) is the j-th column effect, and t is the typical value. But since the response y(i,j) can be expressed as

$$y(i,j) = t + r(i) + c(j) + z(i,j)$$

we have

$$\begin{aligned} y(i,j) &= t + r(i) + c(j) + r(i)*c(j)/t \\ &= t*\{1 + r(i)/t\}*\{1 + c(j)/t\} \end{aligned}$$

Taking logarithms, it follows that

$$\begin{aligned} \log\{y(i,j)\} &= \log(t)+\log\{1+r(i)/t\}+\log\{1+c(j)/t\} \\ &= T + R(i) + C(j) \end{aligned}$$

These manipulations show that if the plot of the residuals z(i,j) versus the comparison values r(i)*c(j)/t follows a line with unit slope, then it is appropriate to reexpress the data in logarithms before doing the median polish. On the other hand, if the fitted line has slope close to zero, then no reexpression is indicated, since the residual plot has no linear structure. We may infer that a tilt in the residual plot having a slope between zero and one suggests a transformation less extreme than logarithms, such as a square root reexpression, while tilts having slopes greater than one suggest reciprocal power reexpressions.

The plot of residuals from a row-plus-column relation versus comparison values is called a diagnostic plot of residuals. For ease of computation, suppose that the program "medpolish" produces both residuals and comparison values if the parameter "resids" is set to the value 2. We can then get the diagnostic plot by invoking the program "scat" on the resultant array. For the consumption data, we have

```
medpolish consumption {resids=2} > z
scat z
```

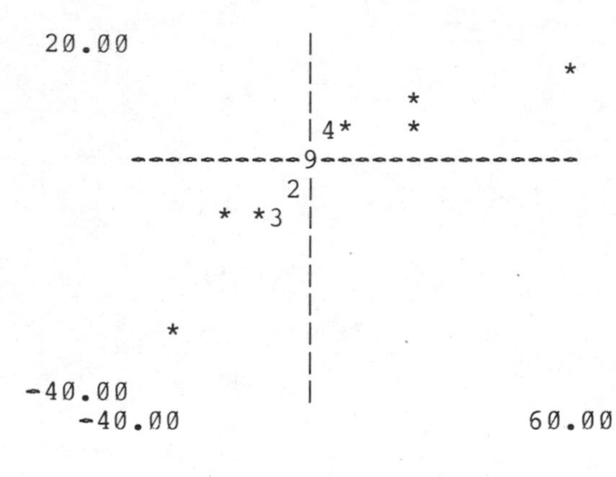

The diagnostic plot shows a clear tilt. To measure the slope, we can use the program "line:"

```
line z

slope:        1.05321   y-intercept:        0.00000
```

One advantage of taking logarithms of data given as a two-way table, rather than using other reexpressions, is that the row-plus-column relation is turned into a row-times-column relation ("log tables turn adders into multipliers") which is just as easy to interpret. There are situations, however, when the data are negative and one still wants to fit a row-times-column relation. Such an example arises in the next section. In this case it is difficult to take logarithms, and another method for fitting the multiplicative relation is needed.

To illustrate the versatility of the techniques described here, consider the following array. The data consist of the classifications of women from three populations according to the number of months (measured either from the time of marriage or the termination of the use of contraceptives) that passed before conception. The information is taken from a paper by S.N. Singh, K.C. Chakrabarty, and V.K. Singh ("A Modification of a Continuous Time Model for First Conception, Demography, Vol. 13, 1976, pages 37-44.) The data comprise the results of surveys or censuses, and the populations are (a) Hutterite women, (b) Taichung women, and (c) United States women who participated in the National Fertility Study.

Delay (months)	Hutterites	Taichung	U.S. Study
1	103	375	380
2	53	247	153
3	43	226	94
4	27	201	45
5-6	39	255	93
7-9	27	257	51
10-12	23	177	46
13-24	23	295	68
25+	4	157	28

One question of interest here is the existence of a standard distribution of the waiting time to conception. In an attempt to answer this question using these data, it is reasonable to first convert the counts to cumulative proportions, and then to see if these proportions can be fitted by a row-plus-column relation. The proportions are presumed to be stored under the label "births."

Cut (months)	Hutterites	Taichung	U.S. Study
1	0.302	0.171	0.397
2	0.456	0.284	0.556
3	0.582	0.387	0.654
4	0.661	0.479	0.701
6	0.775	0.595	0.799
9	0.854	0.713	0.852
12	0.921	0.794	0.900
24	0.988	0.928	0.971

Using the median polish procedure, we command

```
  medpolish births
```

and get

```
    0.633
    0.560
    0.548
    0.544
    4.672
row effects:
 -0.374 -0.255 -0.129 -0.050  0.050  0.143  0.210  0.277
column effects:
    0.000    -0.166     0.039
typical value:         0.711
```

The diagnostic plot of residuals is obtained by the commands

```
  medpolish births {resids=2} > z
  scat z
```

which yield

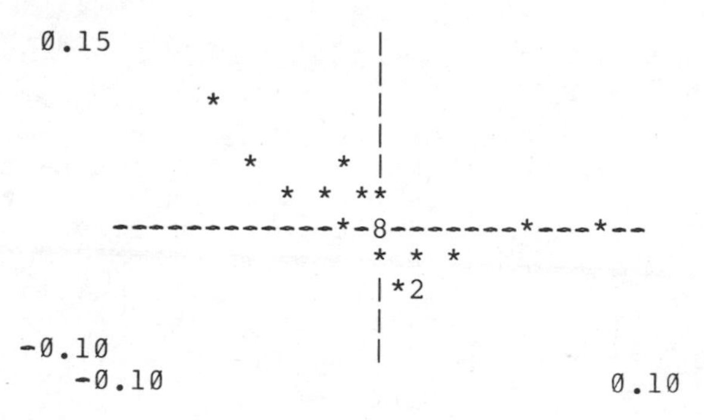

This diagnostic plot shows quite a bit of pattern, but the tilt is opposite to what we would expect if a root or logarithm reexpression were called for. The reason things are not so simple in this case is that the data are proportions, having natural origins at both zero and one, so the class of applicable transformations consists of the folded roots rather than simple powers. Let us try logits:

```
let z=ln(births/(1-births))
medpolish z {resids=2} > zl
scat zl
```

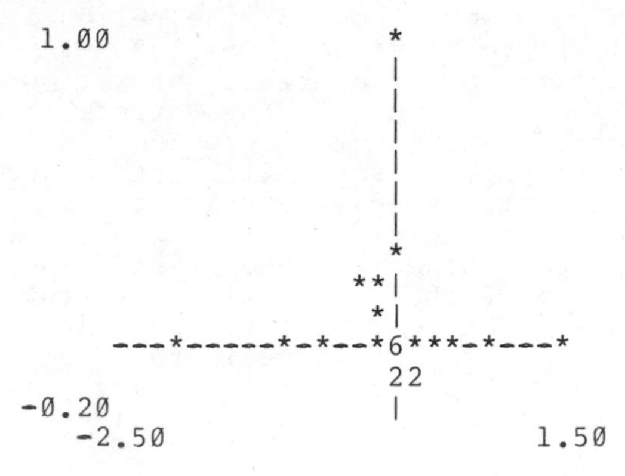

The logit transformation has removed most of the structure from the diagnostic plot, although some resolution from the scatter plot has been lost because of the one very large residual. Let us examine the residuals in more detail, by using the "resids=1" option. Commanding

```
medpolish z {resids=1}
```

we obtain

```
   2.806
   2.402
   2.402
  26.611
row effects:
 -1.715 -1.061 -0.597 -0.264  0.264  0.773  1.213  2.420
column effects:
   0.000    -0.816     0.164
typical value:        0.952

  -0.075     0.000     0.181
  -0.067     0.000     0.170
  -0.025     0.000     0.117
  -0.021     0.043     0.000
   0.021    -0.016     0.000
   0.041     0.000    -0.139
   0.291     0.000    -0.131
   1.039     0.000    -0.025
```

While removing the pattern in the diagnostic plot, the transformation has only slightly increased the variation accounted for from (4.67-0.54)/4.67 = 88% to (26.6-2.8)/26.6 = 89%. However most of the residual sum is contributed by the large residual, 1.039, in the bottom left-hand corner of the table of residuals. This corresponds to the last count in the Hutterite population (4) and probably should be discounted since the number of cases in this cell is so small that a relatively large sampling error can easily result.

Examining the values of the residuals, we see that there is in fact quite a lot of structure remaining - values are negative in the top left-hand and bottom right-hand corners and positive in the other two corners. However there does not seem to be any simple folded transformation for which this residual pattern is satisfactorily removed.

If we accept this relation as the best that can be fitted to the data, we have a formula for the logit of the waiting time distribution of the form

$$\log[p(i,j)/\{1-p(i,j)\}] = t + r(i) + c(j)$$

where t is the typical value (0.952), r(i) are row (month) effects, and c(j) are column (population) effects (0.000, -0.816, and 0.164). Inverting this formula, we obtain

$$p(i,j) = 1/[1 + \exp\{-t-r(i)-c(j)\}]$$

which describes how the proportion who have conceived increases. The Hutterite population, whose column effect is zero, may be regarded as generating a standard schedule of waiting time to conception, and it is of interest to examine this schedule. Let us suppose that the values of t+r(i) are stored in the second column of a data array labeled "months," with the month delays in the first column. Then we generate the standard waiting time distribution using the program "let":

```
let z=1/(1+exp(-months)) {col=2}
```

We can now graph this distribution against month using

the command

```
scat z
```

from which we obtain

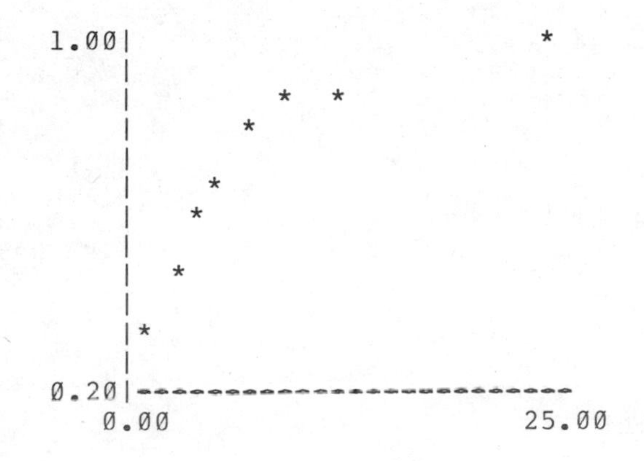

While more resolution is needed, there appears to be a dip in the curve at about 12 months, which may suggest further examination.

HIGHER-ORDER FITS

Fitting a row-plus-column or a row-times-column relation to a two-way table involves choosing one constant, or parameter, for each row and one for each column. The test of how well the relation summarizes the data in the table involves examining the residuals. In some cases no matter how we reexpress the data, a satisfactory fit can not be obtained - there is still structure left in the residual plots. In these cases it may be necessary to fit a more complex relation, one that involves more than one parameter for each row and column.

As an illustration consider the following data, comprising the numbers of telephones in various continents of the world. (Source - The World's Telephones, 1961, American Telephone and Telegraph Company, pages 2 and 3). The counts are given in thousands.

Year	N.Am.	Eur.	Asia	S.Am.	Ocean.	Afr.	Mid.Am.
1951	45939	21574	2876	1815	1646	895	555
1956	60423	29990	4708	2568	2366	1411	733
1957	64721	32510	5230	2695	2526	1546	773
1958	68484	35218	6662	2845	2691	1663	836
1959	71799	37598	6856	3000	2868	1769	911
1960	76036	40341	8220	3145	3054	1905	1008
1961	79831	43173	9053	3338	3224	2005	1076

Suppose the data are stored in an array labeled "phones." To carry out a median polish, and display the diagnostic plot of residuals, we would use the following commands and obtain the responses shown:

```
  medpolish phones {resids=2} > z

 99770.000
 99770.000
719145.000
row effects:
 -1045.0  -325.0  -165.0     0.0   177.0   363.0   533.0
column effects:
 65639.0 32373.0  3817.0     0.0  -154.0 -1182.0 -2009.0
typical value:    2845.00

  scat z
```

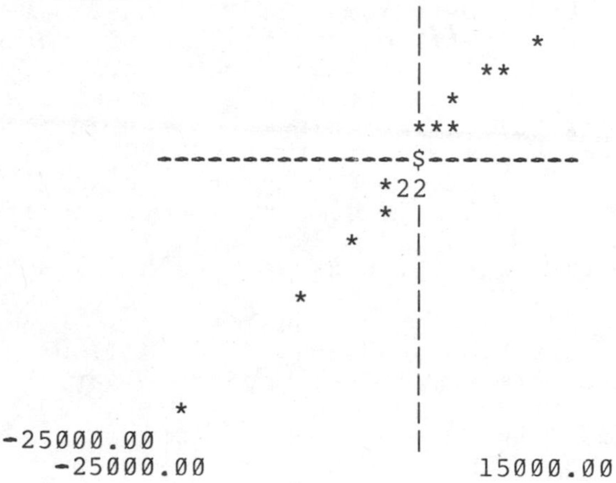

```
line z

slope:       1.24787   y-intercept:        0.00000
```

Clearly, a logarithmic reexpression is indicated, since the diagnostic plot shows residuals congregating along a line whose slope is close to one. The analysis with the logarithms proceeds as follows:

```
  let z=log(phones)
  medpolish z {resids=2} > zl

   0.919
   0.744
   0.744
  25.704
row effects:
  -0.213  -0.070  -0.032   0.000   0.023   0.059   0.081
column effects:
   1.382   1.093   0.359   0.000  -0.024  -0.233  -0.518
typical value:        3.454

  scat zl
```

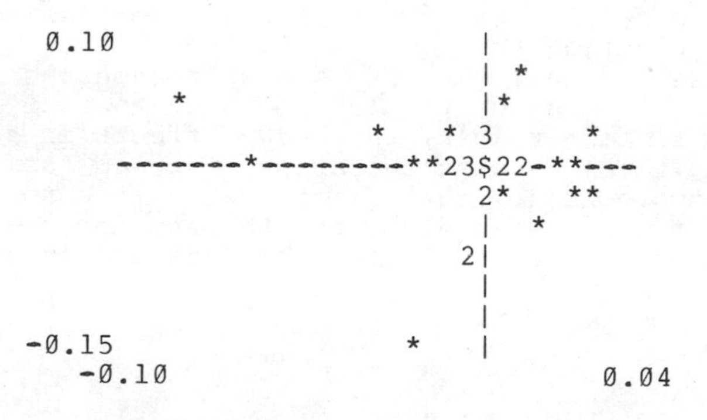

Although this diagnostic plot of residuals appears to be random, there are other ways of looking at residuals which may show pattern. Consider, for example, schematic comparison plots of the columns of the array of residuals:

```
  medpolish z {resids=1} > zl
  compare zl
```

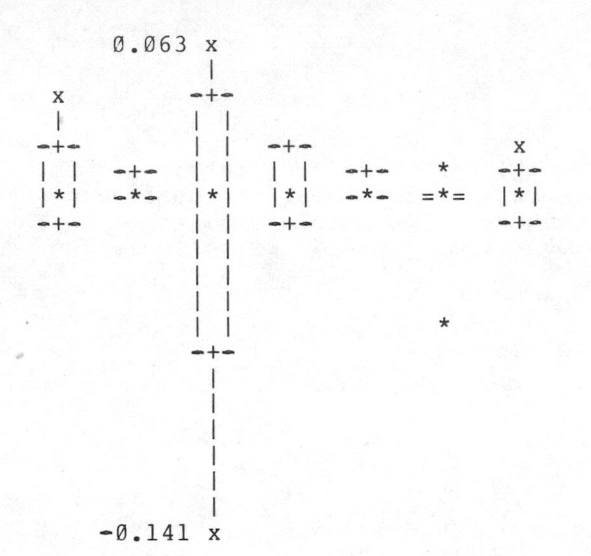

The third column of residuals, corresponding to Asia, has a much greater spread than those for the other continents. Let us examine these residuals more closely. Recall that in the case of fitting a straight line relation to pairs of (x,y) points, we examined residuals by plotting them against fitted values. In the present case, by analogy, it seems reasonable to plot residuals for a particular column against the corresponding fitted values, the row effects in this case. Suppose, then, we plot the array, labeled "asia," consisting of two columns, the first being the row or year effects from the median polish and the second column comprising the residuals for the Asia column.

```
scat asia

  Ø.Ø75                    |     *
                           |    *
                           |
      ---------------------*-*------
                           |
                           |
                     *  *  |
                           |
                           |
 -Ø.15   *                 |
   -Ø.25                         Ø.1Ø
```

The residuals tend to congregrate around a straight line. The parameters in the fitted line are found using the program "line" to be

slope: 0.87597 y-intercept: -0.00868

Most of the pattern in the residuals can be removed by fitting an additional relation for the Asian continent. The y-intercept in the fitted straight line is so close to zero that we may as well put it equal to zero, for simplicity, in which case it only requires one additional constant, 0.876, to improve the overall fit of the row-plus-column relation to the data. In algebraic terms, if y(i,j) represents the logarithms of the telephone counts, we can write the augmented relation as

$$y(i,j) = t + r(i) + c(j) + C(j)*r(i)$$

where t is the typical value (3.454), r(i) are the row or year effects (-0.213, -0.070, -0.032, 0, 0.023, 0.059, and 0.081), c(j) are the column or continent effects (1.382, 1.093, 0.359, 0, -0.024, -0.233, and -0.518), and C(j) is zero unless j=3, when it takes the value 0.976. It may be possible to make further improvements in the goodness of fit by introducing additional constants for the other continents; this would be equivalent to replacing the zeros in the array C(j) by fitted constants. The fitting could be accomplished by displaying scatter plots of the residuals for each continent against the year effects, and fitting straight lines to these plots.

The relation contains one parameter for each year and two parameters for each continent. In algebraic terms it may be written more concisely as

$$y(i,j) = t + c1(j) + r(i)*c2(j)$$

and it may be referred to as a "column-plus-row-times-column" relation. (If the rows and columns were reversed, of course, it would be a "row-plus-row-times-column" relation.) More generally, we could have relations with two parameters for each row and two parameters for each column. In algebraic terms two such "higher-order" fits would be

$$y(i,j) = t + r(i) + c(j) + R(i)*C(j)$$

and

$$y(i,j) = r(i)*c(j) + R(i)*C(j)$$

Higher-order relations are more difficult to fit to two-way tables than row-plus-column fits since there are more parameters, and more iterations are required to obtain a satisfactory fit. However if we have a way of fitting a row-times-column relation, we can fit higher-order relations sequentially, by first fitting one of the component relations to the data, and then fitting the other component to the residuals from the first fit.

EXERCISES

1. Fit a row-plus-column relation to the data on Prussian infant mortality in Exercise 4 of Chapter 2. How much of the variation is explained by this relation? Graph the year effects against time.

2. Do the same with the illiteracy rates in Exercise 6 of Chapter 2. By displaying and examining the diagnostic plot of residuals, determine whether the row-plus-column relation can be improved by a transformation of the data. Discuss the conclusions from the analysis. Can you suggest any alternative fitting procedures that might be more appropriate with these data?

3. By performing a two-way analysis of the election data given in Exercise 2 of Chapter 3, suggest an improved procedure for election forecasting over that which emerged from the one-way analysis. How much better can one do by fitting state effects as well as election-year effects? Comment on any particular features that emerge from these data, and relate them, if possible to knowledge about the candidates.

4. Perform a two-way analysis on any of the other two-factor data arrays listed in the exercises in preceding chapters, using where appropriate a transformation of the data to improve the fit of the

relation.

5. The following data are the world record times in seconds in men's distance events in various past years. (Source - The World Almanac and Book of Facts, 1925, 1945, 1955, 1965, 1971, and 1976.) Fit a row-plus-column relation to these data. Using a diagnostic plot of residuals, search for a reexpression of the data for which the fit is optimal in some sense. By examining the residuals from the improved relation, suggest which events are likely to have records broken, and give estimates for the times that will be recorded when these records are broken. How could you test the efficacy of your forecasting procedure by further examination of the data?

Year	100yds	220yds	440yds	880yds	1mile	2miles	10miles
1924	9.6	20.8	47.4	112.2	250.4	549.6	3040.6
1938	9.4	20.3	46.4	109.6	246.4	536.0	3015.0
1954	9.3	20.2	46.0	108.6	238.0	520.4	2892.0
1964	9.2	20.0	44.9	105.1	234.4	509.6	2867.0
1970	9.1	19.5	44.7	104.9	231.1	499.8	2822.2
1975	9.0	19.5	44.5	103.0	229.4	493.8	2764.2

6. The following data were collected from family records by Richard E. Forrestel, Jr., a student at Princeton University, and analyzed in a term paper for Statistics 101, taught by Ms. J.A. Menken in Fall 1975. The data are the birth weights in pounds and the heights in inches for 10 children in the family at various ages. (Father's height = 5'10", mother's height = 5'5".) Analyze these data and summarize your conclusions.

Age	Steve	Anne	Pete	Mary	Ryan	Andy	Ric.	Col.	Julie	Sara
5	42.	40.5	41.2	41.2	42.	42.5	45.2	42.5	42.8	41.8
6	44.8	43.2	44.2	43.5	43.8	44.8	48.	45.8	46.	44.
7	47.2	46.2	46.8	45.5	46.8	47.8	51.2	47.5	47.5	46.
8	49.2	48.2	49.2	47.5	48.8	49.8	54.	49.5	50.	48.2
9	51.5	50.	51.	50.5	52.8	52.8	56.5	51.8	52.	50.8
10	53.	52.5	53.	51.8	53.5	54.8	58.5	53.8	54.8	52.8
11	55.2	54.	55.	54.	55.2	56.2	61.	55.8	56.5	54.8
12	57.	56.2	56.	55.8	57.8	59.5	62.8	57.5	58.2	56.8
14	60.8	60.	60.2	60.	61.	64.	69.	63.	64.	59.8
Mat.	72.	65.	69.	62.2	71.8	72.8	74.	64.5	66.	61.8
B.Wt.	9-13	8-8	9-12	8-4	9-1	9-6	9-11	9-11	7-10	8-10

7. Collect some data that interest you and analyze them according to the methods suggested in Chapter 5.

PROGRAM COMPONENTS

This section contains listings of the program "medpolish" referred to in the text. There are two parameters, "resids" ("iresids" in FORTRAN), and "epsilon." The parameter "resids" produces residuals if set to 1, comparison values and residuals in a two-column array if set to 2, and does not produce residuals if set to Ø, which is the default value used in the text. The second parameter, "epsilon," is the proportion by which the sum of the absolute values of the residuals must be reduced at each iteration for the program to continue polishing. The default value of "epsilon" used in the text is Ø.Ø1.

APL FUNCTION

In this function the typical value, row effects, and column effects are computed and stored, but not printed out. They are kept under the global variable labels TV, RE, and CE, respectively. These effects are normalized differently from what is stated in the text - TV is the median of the data array.

```
        ∇MEDPOLISH[⎕]∇
        ∇ Z←MEDPOLISH X;R;C;N;RI;CI;RA;CA;S;RG
[1]     R←⍴X[;1]
[2]     N←R×C←⍴X[1;]
[3]     X←X-TV←0.5×Z[⌊0.5×N+1]+(Z←Z[⍋Z←,X])[⌈0.5×N+1]
[4]     RE←R⍴0
[5]     CE←C⍴0
[6]     ⎕←S←+/|,X
[7]     Z←(R,C)⍴(,X)[⍋,X+(⍳R)∘.×C⍴1×RG←(⌈/,X)-⌊/,X]
[8]     X←X-(RI←+/0.5×Z[;⌊0.5×C+1]+Z[;⌈0.5×C+1])∘.×C⍴1
[9]     RE←RE+RI
[10]    Z←⍉(C,R)⍴(,⍉X)[⍋,⍉X+(R⍴1)∘.×(⍳C)×RG]
[11]    X←X-(R⍴1)∘.×CI←+⌿0.5×Z[⌊0.5×R+1;]+Z[⌈0.5×R+1;]
[12]    CE←CE+CI
[13]    →6×⍳EPSILON<1-+/|,X÷S
[14]    Z←X
[15]    →0×⍳RESIDS≠2
[16]    Z←(N,2)⍴((,RE∘.×CE)÷TV),,Z
        ∇
```

FORTRAN SUBROUTINES

The data are stored in the array x, having nr rows and nc columns. Residuals are stored in an array z of similar dimension. r and c are arrays of size nr and nc, respectively, containing the row and column effects, while y is an array of size n=nr*nc used for temporary storage. The subroutine "polish" is used to carry out each iteration.

```
        subroutine polish(z,y,r,c,t,n,nr,nc,sum)
        dimension z(nr,nc),y(n),r(nr),c(nc)
        sum=Ø
        do 3 i=1,nr
        do 1 j=1,nc
1       y(j)=z(i,j)
        call sort(y,nc)
        rdelta=Ø.5*(y(int(Ø.99+nc/2))+y(int(1.Ø+nc/2)))
        do 2 j=1,nc
2       z(i,j)=z(i,j)-rdelta
3       r(i)=r(i)+rdelta
        do 4 j=1,nc
4       y(j)=c(j)
        call sort(y,nc)
        delta=Ø.5*(y(int(Ø.99+nc/2))+y(int(1.Ø+nc/2)))
        do 5 j=1,nc
5       c(j)=c(j)-delta
        t=t+delta
        do 8 j=1,nc
        do 6 i=1,nr
6       y(i)=z(i,j)
        call sort(y,nr)
        cdelta=Ø.5*(y(int(Ø.99+nr/2))+y(int(1.Ø+nr/2)))
        do 7 i=1,nr
        z(i,j)=z(i,j)-cdelta
7       sum=sum+abs(z(i,j))
8       c(j)=c(j)+cdelta
        do 9 i=1,nr
9       y(i)=r(i)
        call sort(y,nr)
        delta=Ø.5*(y(int(Ø.99+nr/2))+y(int(1.Ø+nr/2)))
        do 1Ø i=1,nr
1Ø      r(i)=r(i)-delta
        t=t+delta
        return
        end
```

```
      subroutine medpolish(x,y,z,r,c,n,nr,nc,iresids,
      +epsilon)
      dimension x(nr,nc),z(nr,nc),y(n),r(nr),c(nc)
      do 1 i=1,nr
1     r(i)=Ø
      do 3 j=1,nc
      do 2 i=1,nr
2     z(i,j)=x(i,j)
3     c(j)=Ø
      t=Ø
      oldsum=Ø
      call setfil(11,'/dev/tty')
4     call polish(z,y,r,c,t,n,nr,nc,sum)
      write(11,1Ø) sum
      if(abs(1.Ø-oldsum/sum).le.epsilon) goto 6
      oldsum=sum
      goto 4
6     do 7 i=1,nr
      do 7 j=1,nc
7     y(j+nc*(i-1))=x(i,j)
      call sort(y,n)
      amed=Ø.5*(y(int(Ø.5*n+Ø.99))+y(int(Ø.5*n+1.Ø)))
      sum=Ø
      do 8 i=1,n
8     sum=sum+abs(y(i)-amed)
      write(11,1Ø) sum
      write(11,9)
9     format('row effects:')
      write(11,1Ø) (r(i),i=1,nr)
1Ø    format(15f1Ø.3)
      write(11,11)
11    format('column effects:')
      write(11,1Ø) (c(j),j=1,nc)
      write(11,12) t
12    format('typical value:   ',f1Ø.3)
      if(iresids.eq.Ø) return
      if(iresids.eq.2) goto 14
      do 13 i=1,nr
13    write(6,1Ø) (z(i,j),j=1,nc)
      return
14    do 15 i=1,nr
      do 15 j=1,nc
15    write(6,1Ø) (r(i)*c(j)/t,z(i,j))
      return
      end
```

CHAPTER 6

SMOOTHING

Multiplication is vexation,
Division is as bad;
The rule of three doth puzzle me,
And practice drives me mad.

Anonymous Elizabethan ms. (1570)

SMOOTHING SEQUENCES

In this chapter we develop further the topic of fitting a relation to pairs of (x,y) points, an idea that was introduced in Chapter 3. There we considered a linear relation of the form

$$y = a + b*x$$

Although a linear relation is the simplest that can be fitted to (x,y) pairs, there are times when it is desirable to fit a more general relation. It is true that the straight line relation leads to greater generality when one allows transformations of one or both variables, but in many situations even this degree of generality is insufficient. Suppose, for example, we are interested in forecasting the weather, or in discovering if the temperature in the Northern Hemisphere is changing systematically with time, on the average over a number of decades. It is unlikely that such climatological series can be fitted adequately by a straight line or any mathematical formula derived from a linear relation. Yet what is needed is a relation which describes the systematic structure in the data, so that slowly varying components can be separated from less predictable residuals.

Thus it is desirable to have a procedure for _free smoothing_ - a procedure that does not needlessly constrain the shape of the fitted relation. In Chapter 5 we did something similar to free smoothing when we fitted relations to two-way tables: the relations were determined by the data, the only constraint being one of additivity of row and column effects. In that case, however, an extra dimension was available, so that the relation could be fitted by a kind of "swinging" from rows to columns. [In geometric terms what we were doing was smoothing a surface of responses, since the responses (the data) can be taken as the vertical coordinates and the row and column indices as horizontal coordinates.]

When smoothing (x,y) pairs the extra dimension in the geometric representation for two-way tables is not available, and some other approach is needed. One procedure would be to order the points according to their x-values, and to smooth the responses (the y-values) by replacing each response by a typical value based on responses at adjacent x-values. In this section we consider a very simple, very effective, and very old approach - the use of running medians of three responses to smooth the data. To simplify matters while still covering a wide area of application, it is assumed that the x-values are equispaced. (It may be of interest to think about how the methods that follow may be extended to the case of x-values which are not equispaced.)

To focus on a practical example, consider the counts of "great" inventions and scientific discoveries in each year from 1860 to 1959, as listed by the World Almanac and Book of Facts, 1975 edition, pages 315 to 318. The data are tabulated, with the counts from 1860 to 1879 in the first row, those from 1880 to 1899 in the second row, etc. Assume they are stored as a single list in the data array "discoveries":

```
  show discoveries

5  3  0  2  0  3  2  3  6  1  2  1  2  1  3  3  3  5 2 4
4  0  2  3  7 12  3 10  9  2  3  7  7  2  3  3  6  2 4 3
5  2  2  4  0  4  2  5  2  3  3  6  5  8  3  6  6  0 5 2
2  2  6  3  4  4  2  2  4  7  5  3  3  0  2  2  2  1 3 4
2  2  1  1  1  2  1  4  4  3  2  1  4  1  1  1  0  0 2 0
```

Invoking the program "scat," stretching the width of the display to fill the page, we obtain a plot of the sequence using the command

```
  scat discoveries {width=51}
```

which yields the graph

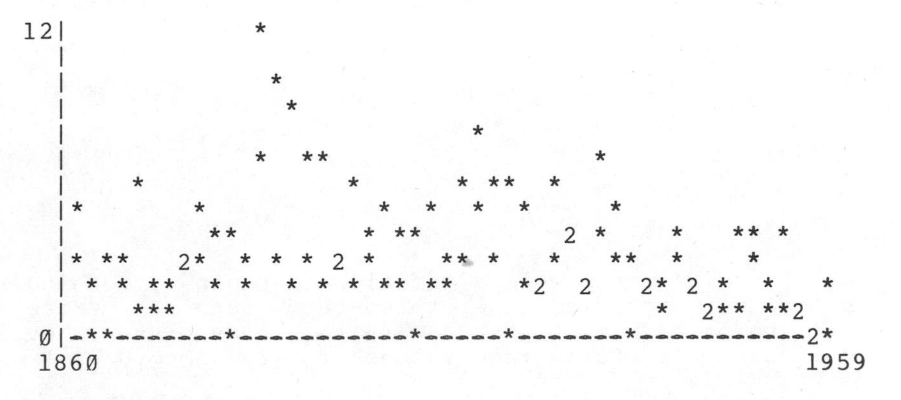

To smooth these data using running medians of 3, we replace the i-th response in the sequence, y(i), by the median of y(i-1), y(i), and y(i+1). The problem of how to treat the end points remains. One <u>end-point rule</u>, suggested by John Tukey in lectures at Princeton University (see his <u>Exploratory Data Analysis</u>), is to replace y(1) by the median of y(1), y(2) and 3*y(2)-2*y(3). The end-point rule for the last point is analogous to this rule.

The rationale behind this end-point rule is to make use of the slope of the line joining the second and third points, in order to smooth the first point. The value 3*y(2)-2*y(3) is actually obtained by fitting a straight line passing through the second point with slope double that of the line joining the second and third points, and taking the value fitted by this relation at the x-value of the first point. The doubling of the slope gives rise to 3*y(2)-2*y(3) rather than 2*y(2)-y(3). The reason for doubling is to allow for the possibility of a greater than linear change in the smoothed curve at the first point. Three cases are shown in Exhibit 3, in which y(2) and y(3) are held constant and y(1) is varied. The smoothed value of y(1) is marked by the symbol o, while 3*y(2)-2*y(3) is given the symbol +.

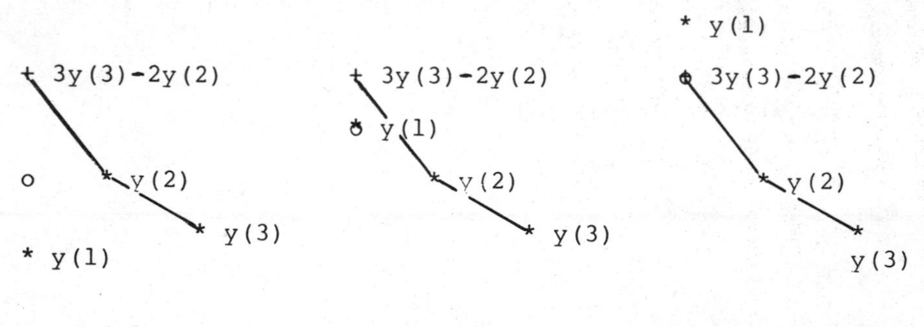

EXHIBIT 3

We are now in a position to begin smoothing the data. Applying the three-term running median smoothing procedure, together with the end-point rule, we obtain the following values for the smoothed sequence:

5	3	2	0	2	2	3	3	3	2	1	2	1	2	3	3	3	3	4	4
4	2	2	3	7	7	10	9	9	3	3	7	7	3	3	3	3	4	3	4
3	2	2	2	4	2	4	2	3	3	3	5	6	5	6	6	6	5	2	2
2	2	3	4	4	4	2	2	4	5	5	3	3	2	2	2	2	2	3	3
2	2	1	1	1	1	2	4	4	3	2	2	1	1	1	1	0	0	0	2

Notice how the end-point rule leaves the first value unchanged at 5, which is the median of 5, 3, and 3*3-2*0 = 9, while the last value is smoothed to 2, which is the median of 0, 2, and 3*2-2*0 = 6.

Here, as elsewhere, there are gains to be made by repeating the fitting procedure. Iterating the preceding smoothing algorithm until no further changes occur, we obtain the values

5	3	2	2	2	2	3	3	3	2	2	2	2	2	3	3	3	3	4	4
4	2	2	3	7	7	9	9	9	3	3	7	7	3	3	3	3	3	3	3
3	2	2	2	2	2	3	3	3	3	3	5	5	6	6	6	6	5	2	2
2	2	3	4	4	4	2	2	4	5	5	3	3	2	2	2	2	2	3	3
2	2	1	1	1	1	2	4	4	3	2	2	1	1	1	1	0	0	0	0

The effect of the iteration is not great, but it is worth doing, particularly if a computer is doing the calculations. In any case, there is not a great amount of effort required if one is using a pencil and paper, since a relatively small number of changes need to be made after the first step.

This procedure can be taken further. Use of medians (rather than means) tends to create mesas, or pairs of adjacent points with a common value which is below or above the points on each side. (A mesa is thus a flat two-point local maximum or minimum.) It is usually desirable to smooth these mesas, and a convenient way of doing so is to divide them, and to apply the end-point rule separately to the values on each side of the division. We can now go back and take running medians of three, repeating until convergence. This procedure - use end-point rule to divide mesas and reapply running medians to the result until convergence - is called <u>splitting</u>. In the present example there are seven mesas, whose values are 2, 3, 7, 2, 5, 3, and 4, respectively (underscored in the preceding display). The sequence that results from splitting the smoothed data is as follows:

```
5  3  2  2  2  2  3  3  3  2  2  2  2  2  3  3  3  3 4 4
4  4  3  3  7  7  9  9  9  9  7  3  3  3  3  3  3  3 3 3
3  2  2  2  2  2  3  3  3  3  3  5  5  6  6  6  6  5 2 2
2  2  3  4  4  4  4  4  4  4  3  3  3  2  2  2  2  2 2 2
2  2  1  1  1  1  2  4  4  3  2  1  1  1  1  1  Ø  Ø Ø Ø
```

Observe that there are still two mesas left - the 23rd and 24th values (3), and the 88th and 89th (4). The first mesa, but not the second one, disappears when the splitting is repeated. With our present tools - medians of 3 and the end-point rule - we have smoothed the sequence as far as we can, and the final smoothed sequence is

```
5  3  2  2  2  2  3  3  3  2  2  2  2  2  3  3  3  3 4 4
4  4  4  7  7  7  9  9  9  9  7  3  3  3  3  3  3  3 3 3
3  2  2  2  2  2  3  3  3  3  3  5  5  6  6  6  6  5 2 2
2  2  3  4  4  4  4  4  4  4  3  3  3  2  2  2  2  2 2 2
2  2  1  1  1  1  2  4  4  3  2  2  1  1  1  1  Ø  Ø Ø Ø
```

The whole smoothing operation we have used may be regarded as a gradual erosion of the roughnesses in the sequence by repeated small steps. The procedure is quite effective, perhaps too effective, since it is actually possible for data behavior to be propagated from one end of the sequence to the other by means of the repeated splitting. To avoid this possibility, some prefer to repeat the splitting once only. This alternative seems to raise another problem, however, since the result of smoothing a sequence would then depend on how much it had been smoothed beforehand.

While the procedure is fairly simple to describe, it can become rather tedious in practice, and it is useful to have a computer to do the work. Suppose we have a program called "smooth3RSR" for routinely performing the smoothing. ("3RSR" is an acronym for "medians of 3, repeated until convergence, split, repeated until convergence.") To obtain a plot of the smoothed sequence, we command

```
smooth3RSR discoveries > z
scat z {width=51}
```

and obtain the response

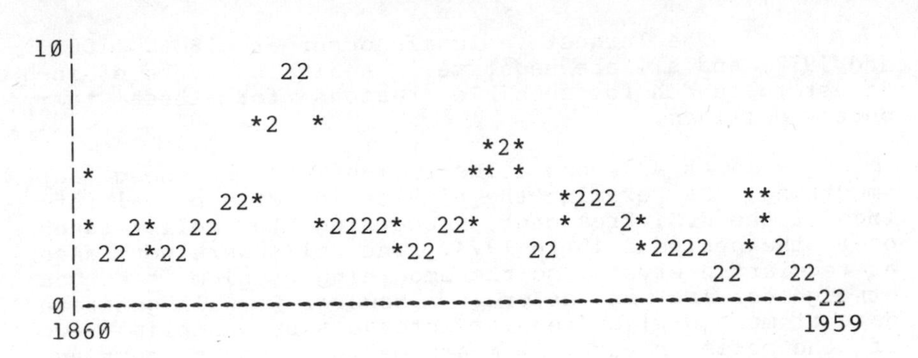

The pattern in this graph is considerably smoother than that in the original data - it is now a simple matter to trace a smooth curve through the points. The graph reveals several peaks that may be of interest to scientific historians, particularly that representing the quite prolonged period from 1885 to 1894. (Curiously, discoveries seem to have increased during World War I, but decreased during World War II.

It is also of interest to look at the residuals after the smoothed sequence has been subtracted from the data. Using the program "let," we proceed with the analysis as follows:

```
let zl = discoveries-z
scat zl {width=51}
```

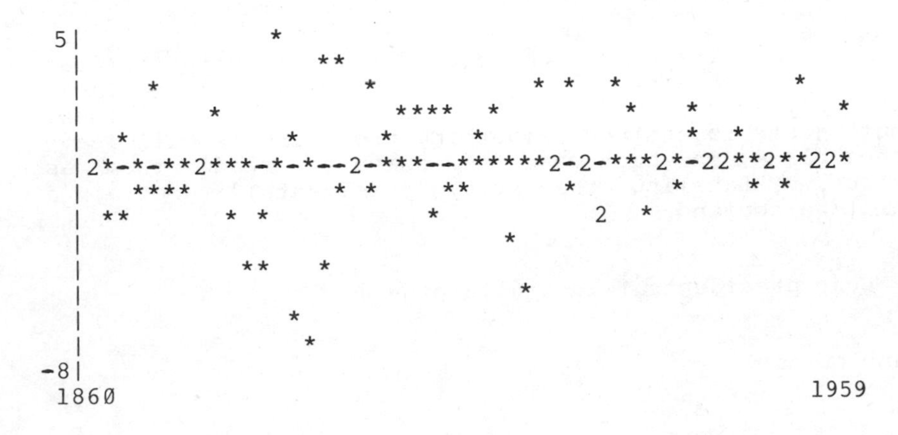

The largest residuals occur at 1886, 1889, and 1917, and all are negative. Again it may be of interest to search for possible reasons for these temporary declines.

As a second illustration of the need for smoothing, we examine the time series of approval ratings of the U.S. President, according to polls taken over the period 1945 to 1974. The polls were not taken at regular intervals, so the smoothing problem is not as straightforward as before. However we will generate data at more or less regular intervals by sampling two of the polls in each six-month period - the highest and the lowest. These data are as follows: (Source - The Gallup Organization, Princeton, N.J.)

1945-46		Jn 87	No 82	No 75	Fe 63	Ap 50	Jy 43	Oc 32
1947-48	Ja 35	Mr 60	Jy 54	Oc 55	Ap 36	Jn 39		
1949-50	Ja 69	Mr 57	Jy 57	Oc 51	Fe 45	Jn 37	Jy 46	Oc 39
1951-52	Ja 36	Jn 24	Oc 32	No 23	Fe 25	Jn 32		No 32
1953-54	Mr 59	My 74	Se 75	De 60	Fe 71	Jn 61	Au 71	No 57
1955-56	Mr 71	My 68	Au 79	Se 73	Mr 76	My 71	Au 67	De 75
1957-58	Fe 79	Jn 62	Jy 63	No 57	Ja 60	Ap 49	Au 48	No 52
1959-60	Ja 57	Jn 62	Au 61	No 66	Ja 71	My 62	Jy 61	Jy 57
1961-62	Mr 72	Ap 83	Jy 71	De 78	Mr 79	Jn 71	Oc 62	De 74
1963-64	Ja 76	My 64	Se 62	Oc 57	Ja 80	Mr 73	No 69	De 69
1965-66	Ja 71	My 64	Jy 69	De 62	Ja 63	Jn 46	Jy 56	No 44
1967-68	Jn 44	Jn 52	Oc 38	De 46	Mr 36	Ap 49	Au 35	De 44
1969-70	Ja 59	My 65	Jy 65	Oc 56	Fe 66	Mr 53	Jy 61	De 52
1971-72	Fe 51	Jn 48	Oc 54	De 49	Ja 49	My 61		
1973-74	Ja 68	Jn 44	Jy 40	No 27	Fe 28	My 25	Jy 24	Au 24

The blanks correspond to six-month periods during which fewer than two polls were taken. These probably should be filled in with interpolated values, but just to test our smoothing procedure we will leave them blank, in which case the computer reads them as zeros. Labeling the array "presidents," to obtain a plot we command

```
scat presidents {width=61,depth=21}
```

and obtain

```
100|
   |
   |*                               *
   | *                   *  *        **   *
   | *              **   ***           ** *
   |        *         **2 *       * ***    2**              *
   |  *                    * *   *      *   * *     ***
   |    *           ***     * * ****   * *   *      *  *  *
   |     2  ** *       *     *  *  *     *     *     **  *
   |  *      * *              *2                * *    *2**
   |   *      *                               **** *        *
   |      *                                      *           *
   |    * *   * *                                 **
   |   *         ***                                          *
   |            ***                                          **2
   |
   |
   |
   |
  0|*------2-------*---------------------------------------2----
  1945                                                      1975
```

To obtain a plot of the smoothed sequence, we command

```
smooth3RSS presidents > z
scat z {width=61}
```

and obtain

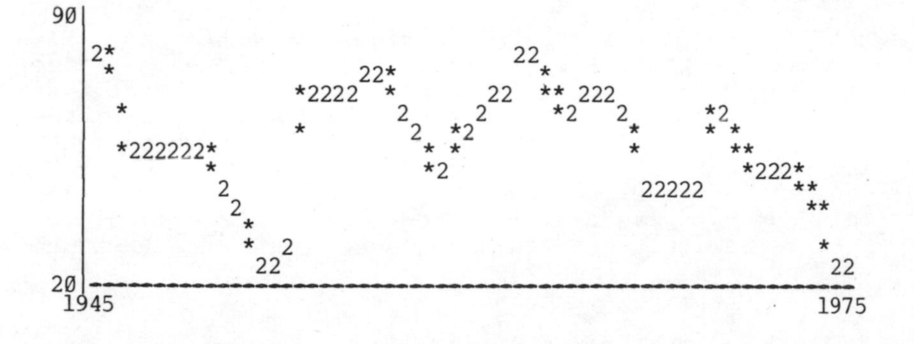

Several features show up quite clearly in the smoothed sequence - Truman's two slides, Eisenhower's heart attack (the first peak in the smoothed sequence) followed by the trough corresponding to the Russian Summit failure, and the Kennedy, Johnson, and Nixon slides. The smoother has also effectively ignored the zeros corresponding to the missing values.

The ability to interpolate missing values is a useful property of a smoothing procedure. Since we have no information concerning these values other than what is contained in the smoothed sequence, we may as well replace the residuals at these points by zeros. Suppose that these adjusted residuals are stored in the array "z." Their plot is obtained as follows:

```
scat z {width=61}
```

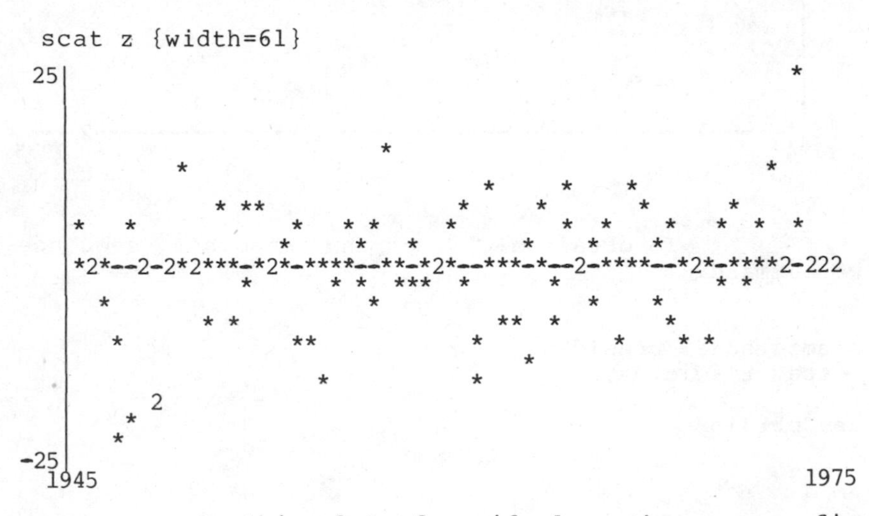

In this plot of residuals, there are five outliers, the first four (which are negative) corresponding to Truman's drop in popularity before his election in 1948, and the last one (which is positive) just after Nixon's reelection in 1972. It is interesting to note that in each case the President received a boost in approval rating at the time of the election. In Truman's case this resulted in his unpredicted election to office, while in Nixon's case the result was his reelection by a landslide.

The objective in smoothing, as in any situation where a relation is being fitted to data, is to decompose the data into two parts: the smoothed sequence, or <u>smooth</u>, and the residuals, or <u>rough</u>. As elsewhere in exploratory data analysis, it is not always obvious how to make this decomposition. In the Presidential approval data some would prefer to allocate the responses corresponding to Truman's trough in approval before his election to the smooth, rather than to the rough as we have done. To provide such flexibility, a number of refinements may be made to the basic smoother defined here. Two of these refinements may be described as follows.

The first involves smoothing the rough, and adding the resulting smooth back to the original smooth to obtain the final smooth. This procedure is called <u>twicing</u>. When smoothing the array "discoveries" using "3RSR," twicing has no effect, but there is an effect for the Presidential data. The results of smoothing "presidents" using "3RSR" and "3RSR, twice," respectively, are as follows:

3RSR:

```
82 82 82 75 63 54 54 54 54 54 54 54 54 57 57 57
57 57 57 51 45 45 39 39 36 32 25 25 25 25 32 32
59 71 71 71 71 71 71 71 71 71 73 73 73 73 73 71
65 63 62 60 57 51 51 52 57 61 62 62 64 64 72 72
72 72 78 78 78 78 74 71 68 64 64 64 69 69 69 69
69 69 64 63 62 56 46 44 44 44 44 44 44 44 44 44
59 65 65 65 61 56 53 52 51 51 49 49 49 49 49 44
44 40 40 28 27 25 24 24
```

3RSR, twice:

```
82 82 82 75 63 54 54 54 54 54 54 54 36 39 39 39
57 57 57 51 45 45 39 39 36 32 25 25 25 25 32 32
59 71 71 71 71 71 71 71 71 71 73 73 73 73 73 72
66 64 63 59 55 49 49 52 57 61 63 63 65 64 72 72
72 78 78 78 78 78 74 71 68 64 64 46 69 69 69 69
69 69 64 63 62 56 46 44 44 44 44 44 44 44 44 44
59 65 65 65 61 56 44 52 51 49 49 49 49 49 49 53
44 40 40 28 27 25 24 24
```

In this case the twicing procedure restores some features (underscored) that were washed out by the original smoothing, in particular, Truman's trough.

The second refinement involves using something other than running medians as the basic component of the smoother. In lectures at Princeton University (see <u>Exploratory Data Analysis</u>), John Tukey suggested the alternative of replacing "3" with

$$\underline{3} = \text{median } \{u, v, w\}$$

where

$$u = \text{median } \{y(i-1),\ y(i),\ y(i+1)\}$$
$$v = \text{median } \{y(i-1),\ (3*y(i-1)-y(i-2))/2,\ y(i)\}$$
$$w = \text{median } \{y(i+1),\ (3*y(i+1)-y(i+2))/2,\ y(i)\}$$

The effect of using "$\underline{3}$RSR" rather than "3RSR" is to preserve local peaks in the data to a greater extent. As an illustration consider again the Presidential data array. The result of applying the smoothing operation "$\underline{3}$RSR" is

```
82 82 82 75 63 50 46 48 42 54 54 54 54 57 57 57
57 57 57 51 45 45 39 39 36 32 25 25 25 25 32 32
59 71 71 71 71 71 71 71 71 71 73 73 73 73 71 71
65 63 62 60 57 51 50 52 57 61 62 62 64 64 72 72
72 72 78 78 78 78 74 71 68 64 64 64 69 69 69 69
69 69 64 63 62 56 46 44 44 44 44 44 44 44 44 44
59 65 65 65 61 56 53 52 51 51 49 49 49 49 49 44
44 40 40 28 27 25 24 24
```

Comparing this smooth with the one obtained by using "3RSR," we see that there are only two stretches where the results are different - responses 6-9, and 55. (The 55th value from "$\underline{3}$RSR" is actually 49.5, which has been rounded to the even integer 50.)

Before concluding this section it should be noted that there is one more smoothing operation that is often effective in producing an attractive display, namely, *hanning*. To hann a sequence, the i-th response is replaced by $0.25*y(i-1)+0.5*y(i)+0.25*y(i+1)$. This procedure, named after the Austrian astronomer Julius von Hann, who first suggested it, may be regarded as a final touch-up, similar to sanding and polishing a piece of furniture, which rounds out the corners in the smoothed data.

MULTIPLE REGRESSION

In the preceding section we saw how the concept of fitting a straight line relation to (x,y) pairs could be extended to free smoothing. There is another direction in which the idea of fitting a linear relation can be extended. It often happens that one has to deal with data arrays in which the elements are not just single numbers or pairs of numbers, but groups of three or more numbers - triplets or quadruplets or, more generally, n-tuplets. Several of the data arrays considered before may be looked at in this way.

Consider, for example, the economic GNP data listed in Exercise 4 of Chapter 3. There the question asked was the nature of the relation between private investment and personal consumption, the third component of GNP (government purchases) being ignored. Suppose, however, that we wanted to investigate the relation between all three components. How can this be done?

One approach is to think of the data as a two-way table and to proceed to fit a row-plus-column relation. This would provide a concise summary of the pattern in the data, and some insight could be gained, perhaps, by examining the parameters in the fitted relation. While this approach is certainly worth pursuing, it does not directly face the problem of the relation between the variables. Moreover it treats the variables equally, and we may prefer to single out a particular variable as the response and look at its dependence on the other variables.

If there are only two variables, there is a choice of treating the data as a two-way table with two columns and fitting a row-plus-column relation, or plotting the data as (x,y) points in a scatter plot and fitting a linear relation (or smoothing in some other way, if a less constrained relation is preferred). When there are more than two variables, we still have these choices, but in this case more than two dimensions are needed to plot the data. With three variables the data may be regarded as triplets (x,y,z) and plotted in three-dimensional space, using a straightforward generalization of the scatter plot. In general, the plot of n-tuplets in n-dimensional space may be called a <u>point cloud</u>. As in the case of a scatter plot of pairs of numbers, we must decide which variable is to be plotted vertically; this choice depends on which variable is regarded as the response. To be consistent with the notation for making scatter plots, we will call the response y and the carriers x1, x2, ... , and y will be plotted vertically, as before. (The cases where it is not clear which, if any, variables are carriers and which are responses, are possibly best thought of, at least to begin with, as two-way table situations.)

To focus on a specific problem, consider the

crime rate in the United States. Listed are the statistics, in arrests per 100,000 residents, for assault, murder, and rape in each of the 50 states in 1973. (Source - The World Almanac and Book of Facts, 1975, page 966.) We also give data on percent urban in each state, obtained from the Statistical Abstract of the United States, 1975, published by the U.S. Bureau of the Census (page 20). The data are given as quadruplets, in the order murder, assault, per cent urban, and rape.

```
AL (13.2, 236, 58, 21.2)    AK (10.0, 263, 48, 44.5)
AZ ( 8.1, 294, 80, 31.0)    AR ( 8.8, 190, 50, 19.5)
CA ( 9.0, 276, 91, 40.6)    CO ( 7.9, 204, 78, 38.7)
CT ( 3.3, 110, 77, 11.1)    DE ( 5.9, 238, 72, 15.8)
FA (15.4, 335, 80, 31.9)    GA (17.4, 211, 60, 25.8)
HA ( 5.3,  46, 83, 20.2)    ID ( 2.6, 120, 54, 14.2)
IL (10.4, 249, 83, 24.0)    IN ( 7.2, 113, 65, 21.0)
IO ( 2.2,  56, 57, 11.3)    KS ( 6.0, 115, 66, 18.0)
KY ( 9.7, 109, 52, 16.3)    LA (15.4, 249, 66, 22.2)
ME ( 2.1,  83, 51,  7.8)    MD (11.3, 300, 67, 27.8)
MA ( 4.4, 149, 85, 16.3)    MC (12.1, 255, 74, 35.1)
MN ( 2.7,  72, 66, 14.9)    MS (16.1, 259, 44, 17.1)
MO ( 9.0, 178, 70, 28.2)    MT ( 6.0, 109, 53, 16.4)
NB ( 4.3, 102, 62, 16.5)    NV (12.2, 252, 81, 46.0)
NH ( 2.1,  57, 56,  9.5)    NJ ( 7.4, 159, 89, 18.8)
NM (11.4, 285, 70, 32.1)    NY (11.1, 254, 86, 26.1)
NC (13.0, 337, 45, 16.1)    ND ( 0.8,  45, 44,  7.3)
OH ( 7.3, 120, 75, 21.4)    OK ( 6.6, 151, 68, 20.0)
OR ( 4.9, 159, 67, 29.3)    PA ( 6.3, 106, 72, 14.9)
RI ( 3.4, 174, 87,  8.3)    SC (14.4, 279, 48, 22.5)
SD ( 3.8,  86, 45, 12.8)    TN (13.2, 188, 59, 26.9)
TX (12.7, 201, 80, 25.5)    UT ( 3.2, 120, 80, 22.9)
VT ( 2.2,  48, 32, 11.2)    VA ( 8.5, 156, 63, 20.7)
WA ( 4.0, 145, 73, 26.2)    WV ( 5.7,  81, 39,  9.3)
WI ( 2.6,  53, 66, 10.8)    WY ( 6.8, 161, 60, 15.6)
```

Before we can begin to analyze these data, we really need to have some idea of the question we are hoping to answer. If we are interested in the general relation between crime and urbanization, then perhaps we should first analyze the relation between crime and urbanization separately for each crime. Suppose, however, that our motive for examining these data is that we suspect the rape rate is biased downward in some states because of underreporting, and we want to adjust the reported rate to remove this bias, if possible. In this case we should take the rape rate as the response, and the other variables as carriers. The point cloud of the data is a four-dimensional display which is hard to visualize, but we can get some idea of its shape by looking at two-dimensional projections obtained by plotting the response against each of the carriers in turn. Suppose we have the data stored as a four-column array labeled "crimes." The scatter plots of rape versus murder and rape versus assault are obtained as follows:

```
scat crimes {x=1,y=4}
```

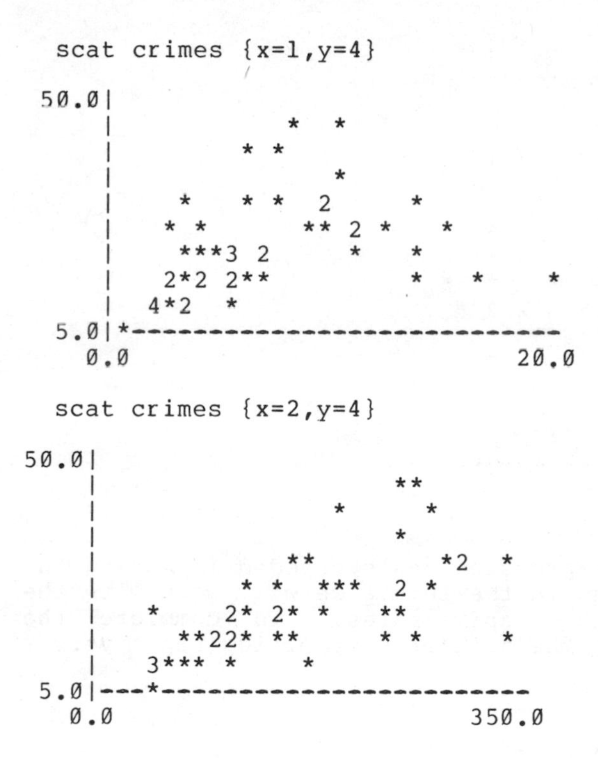

These plots reveal some relation between rape and both murder and assault, as might be expected. The relations in the two cases seem to be about equal in strength. There is, however, some tendency for the points to spread out more as their distances from the origin increase. As in the case of the data array "cars," considered in Section 4 of Chapter 3, this kind of unevenness in spread can usually be removed by a transformation. Let us take logarithms of the crime rates and examine the resulting scatter plots:

```
let logcrimes = log(crimes) {col=1,2,4}
scat logcrimes {x=1,y=4}
```

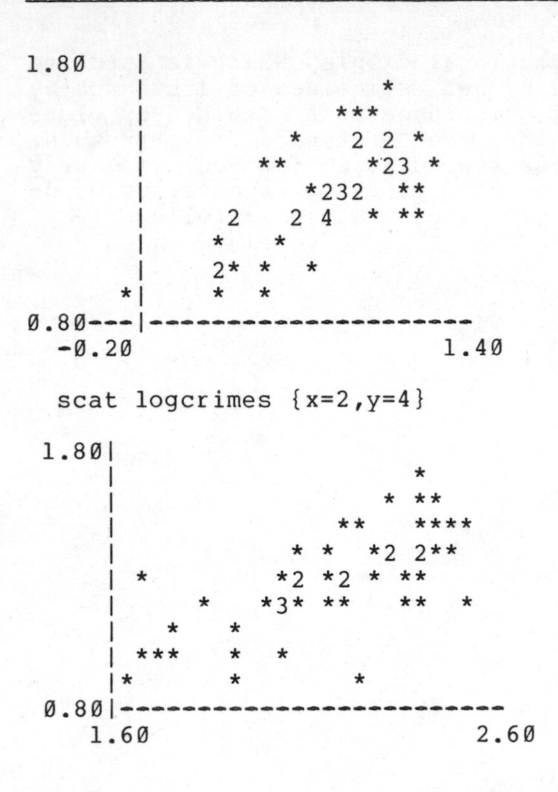

The reexpression has succeeded in making the spread more even, so in the future we will work with the logarithms of the three crime rates. To complete the picture, we give the scatter plot of log(rape) versus percent urban:

scat logcrimes {x=3,y=4}

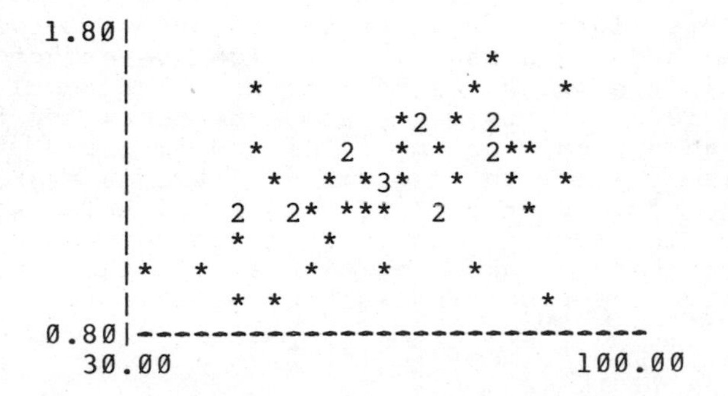

Again there is evidence of a relation between the response and the carrier, though not a particularly strong one. What are we to make of all this? If there were just one carrier, we could fit a straight line to the scatter plot, and use this linear relation to describe what is going on. What we need to do is to reduce the number of variables to two, without throwing away any information. It turns out that there is a procedure for doing this; it is called <u>multiple regression</u>, and may be described as follows:

We begin by ordering the carriers, calling the first x1, the second x2, etc. In the present case, taking the carriers in the order in which they are written down in the quadruplets, we have x1=murder, x2=assault, and x3=urbanization. Next we consider the pairs (x1,x2), (x1,x3), ..., and (x1,y), fitting a straight line to the data in each case. In algebraic terms these relations may be written as

```
x2 = a + b*x1
x3 = c + d*x1
 y = e + f*x1
```

Subtracting the fitted relation from the data in each case, we obtain the residuals

```
x2' = x2 - (a + b*x1)
x3' = x3 - (c + d*x1)
 y' = y - (e + f*x1)
```

We can now use these residuals in place of the original data, and we have one less carrier than when we started. The variable x1 has been eliminated, because its relation with the response and the other carriers has been removed by the subtraction. (Of course this is only strictly true if these relations are linear, which is the assumption made here.)

Having reduced the number of carriers by one, the procedure can now be repeated, and continued until there is only one carrier remaining. The final straight line fitted to this relation gives the dependence of the response on this last carrier when the (linear) effects of all the other carriers have been eliminated by subtraction. Since we are interested, however, in the dependence of the response on <u>all</u> the carriers, we need to go back and piece together the linear relations to get a composite formula. To see how this works, let us go through the algebra for the case of three carriers, and then perform the data analysis on the crime data.

Fitting straight lines to the residual pairs (x2',y') and (x3',y'), we obtain

```
x3' = g + h*x2'
 y' = j + k*x2'
```

Subtracting these fitted relations, we obtain the residuals

```
x3" = x3' - (g + h*x2')
 y" = y' - (j + k*x2')
```

The relation

```
y" = m + n*x3"
```

can now be fitted to this final response-carrier pair. Substituting for y" and x3" we may expand this to

```
y' - j - k*x2' = m + n*(x3' - g - h*x2')
```

Substituting further for y', x2', and x3', we obtain the further expansion

```
y-e-f*x1-k*(x2-a-b*x1)
                  = m + n*{x3-c-d*x1-g-h*(x2-a-b*x1)}
```

that is

```
y = A + B*x1 + C*x2 + D*x3
```

where

D=n, C=k-n*h, B=f-n*d-b*C, and A=e+j+m-n*(c+g)-a*C

Turning to the data array "logcrimes," we can determine the parameters a, b, c, d, e, f, g, h, j, k, m, and n by using the program "line," as follows:

```
  line logcrimes {x=1,y=2}

slope:          0.64202   y-intercept:        1.66292

  line logcrimes {x=1,y=3}

slope:          0.00000   y-intercept:       65.99990

  line logcrimes {x=1,y=4}

slope:          0.46672   y-intercept:        0.89937
```

The parameter values a=1.663, b=0.642, c=66.0, d=0, e=0.899 and f=0.467 thus result. Next we need to subtract the fitted relations; this can be accomplished by the program "let":

```
  let x1 = logcrimes[*,1]
  let x2' = logcrimes[*,2]-1.663-0.642*x1
  let x3' = logcrimes[*,3]-66
  let y' = logcrimes[*,4]-0.899-0.467*x1
  let z = x2',x3',y'
```

Now we fit linear relations to y' versus x2' and y' versus x3', obtaining

```
  line z {x=1,y=2}

slope:          48.82807   y-intercept:         1.01533

  line z {x=1,y=3}

slope:           0.32380   y-intercept:         0.00781
```

Subtracting these fits, and fitting the final relation, we command

```
let x3" = x3'-1.015-48.8*x2'
let y" = y'-0.008-0.324*x2'
let z1 = x3",y"
line z1
```

from which we obtain

slope: 0.00258 y-intercept: 0.00604

The values of the remaining parameters are thus g=1.015, h=48.8, j=0.008, k=0.324, m=0.006 and n=0.0026. Using the formulas, we get D=0.0026, C=0.20, B=0.34, and A=0.41, so the multiple regression is

y = 0.41 + 0.34*x1 + 0.20*x2 + 0.0026*x3

that is

log(rape)
= 0.41+0.34*log(murder)+0.2*log(assault)+0.0026*(%urban).

The coefficients of the carriers in the regression (A, B, C, and D, in the preceding example), are called the <u>regression coefficients</u>. From the relation obtained, it would appear that murder is the most important factor, since its regression coefficient is the largest, and that urbanization is least important, since its coefficient is smallest. This conclusion is not necessarily correct, however, since if the urbanization data were changed to proportions rather than percentages, the corresponding regression coeffient would increase by a factor of 100. What is needed is some way of judging the significance of the various carriers in the regression relation. We can answer this question by examining spreads of residuals.

The residual spreads of interest are those arising after the effect of each additional carrier has been removed in turn. When no carrier has been removed, the midspread of residuals is 0.215, as can be determined using the program "condense":

```
condense logcrimes[*,4]

    1.303     0.215
     med      sprd
```

To compare the spreads of residuals at each step, let us construct an array containing these batches of residuals as columns:

```
let yl = logcrimes[*,4]-1.303
let z = y"-0.006-0.0026*x3"
let zl = yl,y',y",z
```

The routine "compare" may now be used to display the distributions of residuals:

```
compare z
```

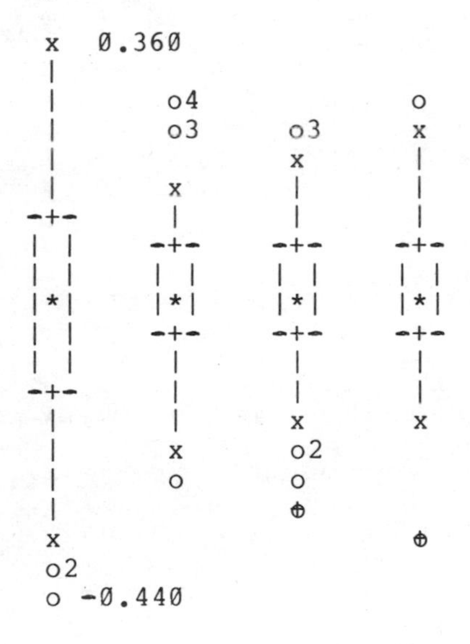

From this display it is apparent that some reduction in residual size occurs after the removal of the effect of the first carrier (murder), but that very little, if any, further reduction occurs when the effects of assault and urbanization are removed. This does not necessarily mean that assault and urbanization are unrelated to rape (in fact our scatter plots indicate that there is some relation), but it does mean that if one is interested in predicting the rape statistic, given the other three statistics, only the first need be considered. (Of course, if there were high correlation between the three carriers, we would expect to do about equally well with any one of them. Recall, however, that the slope of the straight line fitted to the plot of urbanization versus murder was found to be zero, so the situation is not that simple.) In any case we may conclude that the prediction formula leaves something to be desired, since only about half of the spread in the distribution of the rape rate among states can be accounted for by regressing on the three carriers considered.

To conclude our discussion of this example, let us examine the residuals. An effective display may be obtained by locating and marking the residuals for the various states on a map. Suppose we have a program, "usmap," which, given an array of 50 integers corresponding to U.S. states in alphabetical order, produces a map. Since the data are expressed in logarithms, the residuals, when multiplied by 100, are close to the percentage deviations of the rates from those predicted by the regression formula. The results are

```
  let z2 = int(100*zl)
  usmap zl

 29 *******                                                   ***
   *  18  *********************                              *-14*
  *                        4       * **                     ***  *
  *               -1            7**  ** *            * 10 -4
 *  21                                 -3 ** *     ** -5 -7*
 *          5              0              9*   *   -15*-35
*               -10                          ****-10-10*
*                           3      3      4   2     1*-14
*      19                                -7     -18    **
 * 15       15     21              8        -8     -1**
 *                           0                     -20**
  *                                       3           *
   **       7                 0   -3             -7**
     ***          6                       -9  -4 *
        ***                              -18      *
           *******       -5    -12  *******  *
                  ** **        ********     *  *
                    *  *   **               * **
                        * *                  -2*
 10                      **                   *
```

It is not too difficult to discern a regional pattern in this residual map. Sociologists may have some explanation.

There are many more applications of smoothing than can be adequately dealt with in this chapter. One such application - smoothing spatially located data - is suggested by the preceding display.

EXERCISES

1. Smooth the data on British rulers' longevity given in Exercise 1 of Chapter 1. Interpret the smooth and the rough.

2. Do the same with the aircraft fatalities given in Exercise 2 of Chapter 1. State any substantive conclusions that can be made from the analysis.

3. Smooth the New Haven temperature data listed in Exercise 3 of Chapter 1. Is there any evidence of periodicities in the series? If so, how would you examine the data further, in order to fit a periodic relation?

4. For the 10 Forrestel family members whose height growth data are given in Exercise 6 of Chapter 5, regress height at maturity on birth weight and height at five years of age. Discuss the goodness of fit of the components in this regression.

5. Take the GNP data listed in Exercise 4 of Chapter 3, and carry out a regression of personal consumption on private investment and government purchases. Discuss the efficacy of this regression. Plot the residuals, smooth them, and comment on any features of interest.

6. In a national study of energy use, family homes in East Windsor, New Jersey, were continuously monitored, and data were obtained on gas consumption, internal and external temperatures, and external wind velocity. The following are the integrated readings on successive days, starting with November 17, 1975, for one particular house. The data are given as quadruplets, in the following order: internal and external temperatures in degrees Fahrenheit, wind velocity in miles per hour, and gas consumption in cubic feet per hour.

```
(73,54,3.9,0.0),    (73,52,6.8,0.0),    (74,54,10.7,0.8),
(70,41,9.2,8.7),    (73,37,3.6,14.1),   (72,40,9.5,9.6),
(73,40,4.0,9.8),    (72,38,3.8,9.2),    (75,49,8.2,5.0),
(71,40,8.4,11.3),   (72,40,5.9,7.7),    (72,52,8.8,3.2),
(71,60,16.6,7.4),   (72,**,**,14.2),    (73,35,9.5,14.0),
(72,30,4.4,19.2),   (72,40,5.0,9.0),    (72,46,6.3,6.3),
(71,34,9.3,12.6),   (74,31,9.3,22.5),   (75,44,11.6,15.4),
(73,44,9.4,8.4),    (**,38,8.6,**),     (73,40,7.2,13.8),
(74,43,7.4,12.4),   (75,47,5.1,5.9),    (74,52,8.3,3.2),
(73,42,7.9,8.3),    (72,37,4.5,12.8),   (71,28,11.6,19.2),
(72,19,12.3,31.0),  (73,31,6.3,22.1),   (73,25,10.1,22.1),
(73,27,12.8,22.1),  (72,30,10.8,22.7),  (72,**,**,29.0),
(72,**,**,29.4),    (73,48,8.5,11.3),   (74,38,7.0,11.2),
(73,30,8.4,15.3),   (72,29,4.4,15.8),   (72,36,3.7,13.2),
(73,39,5.5,12.4),   (73,33,10.9,16.4),  (72,25,5.0,20.4),
(74,36,6.9,13.0),   (72,27,16.0,21.9),  (72,21,10.3,28.8),
(74,22,3.9,25.5),   (73,29,3.5,20.4),   (73,28,9.3,18.5),
(72,16,11.7,30.0),  (73,18,8.4,32.8),   (74,29,5.4,22.7),
(74,31,7.0,17.4),   (74,34,9.6,17.0),   (72,42,17.4,12.5),
(72,30,10.7,21.5),  (73,34,6.0,16.9),   (74,23,12.9,24.5),
(72,10,12.1,35.2).
```

Asterisks correspond to missing data caused by equipment failure. (Data supplied by courtesy of Dr. Tom Woteki, Assistant Professor in the Statistics Department and Research Statistician in the Center for Environmental Studies, Princeton University.) According to engineering physicists, the heat loss depends on the difference between the external and internal temperatures, and on the product of this difference and the wind velocity. Analyse the data, using as many of the procedures for exploratory data analysis as seem relevant, and comment on your conclusions.

7. Collect some data and analyze them using the procedures outlined in this chapter.

PROGRAM COMPONENTS

The following routines may be used to perform the "3R," "3R," "3RSR," and "3RSR" smoothing operations. In these programs "3" is written as "a3," and "3" as "b3."

APL FUNCTIONS

```
       ∇A3R[⎕]∇
       ∇ Z←A3R X;Y1;Y2;X1;X2;N;C
[1]    Y1←(3×X[2])-2×X[3]
[2]    Y2←(3×X[N-1])-2×X[¯2+N←ρX]
[3]    X1←Y1,¯1↓X
[4]    X2←1↓X,Y2
[5]    Z←(Y1,X,Y2)[(⍳N)+1+(C=X1<X)-(X≤X2)=C←X1<X2]
[6]    →1×0≠+/|(X←Z)-X
       ∇
```

```
       ∇B3R[⎕]∇
       ∇ Z←B3R X;N;J;V;W;Y;C;X0;X1
[1]    X1←(3×X[N-1])-2×X[¯2+N←ρX]
[2]    V←X[2],X[3],(0.5×(3×X[J+1])-X[J←⍳N-3]),X1
[3]    W←(X0←(3×X[2])-2×X[3]),¯1↓X
[4]    Y←1↓X,X1
[5]    Z←X0,(0.5×(3×X[J+2])-X[J+3]),X[N-2],X[N-1]
[6]    V←(X×1=J)+(V×¯1=J)+W×0=J←(C=V<W)-(W≤X)=C←V<X
[7]    W←(Y×1=J)+(W×¯1=J)+X×0=J←(C=W<X)-(X≤Y)=C←W<Y
[8]    Y←(Z×1=J)+(X×¯1=J)+Y×0=J←(C=X<Y)-(Y≤Z)=C←X<Z
[9]    Z←(Y×1=J)+(V×¯1=J)+W×0=J←(C=V<W)-(W≤Y)=C←V<Y
[10]   →1×0≠+/|(X←Z)-X
       ∇
```

```
       ∇SPLIT[⎕]∇
       ∇ Z←SPLIT X;P;Q;R;S;T;N;I;J;C
[1]    P←¯3↓X
[2]    Q←¯2↓1↓X
[3]    R←¯1↓2↓X
[4]    S←3↓X
[5]    I←(((P<Q)×R>S)+(P>Q)×R<S)×Q=R
[6]    Q←(X←X[1],X,X[N])[1+I←(I≠0)/I←I×⍳¯3+N←ρX]
[7]    R←(3×Q)-2×X[I]
[8]    P←X[I+2]
[9]    T←(3×S←X[I+4])-2×X[I+5]
[10]   X[I+2]←(R×1=J)+(P×¯1=J)+Q×0=J←(C=P<Q)-(Q≤R)=C←P<R
[11]   X[I+3]←(T×1=J)+(P×¯1=J)+S×0=J←(C=P<S)-(S≤T)=C←P<T
[12]   Z←1↓¯1↓X
       ∇
```

```
      ∇SMOOTH3RSR[⎕]∇
      ∇ Z←SMOOTH3RSR X
[1]   X←A3R X
[2]   Z←A3R SPLIT X
[3]   →2×0≠+/|(X←Z)-X
      ∇
```

```
      ∇SMOOTHΔRSR[⎕]∇
      ∇ Z←SMOOTHΔRSR X
[1]   X←B3R X
[2]   Z←B3R SPLIT X
[3]   →2×0≠+/|(X←Z)-X
      ∇
```

FORTRAN SUBROUTINES

```
      subroutine a3(x,y,n,d)
      dimension x(n),y(n)
      y(1)=amed3(x(1),x(2),3*x(2)-2*x(3))
      do 1 i=2,n-1
1     y(i)=amed3(x(i-1),x(i),x(i+1))
      y(n)=amed3(x(n),x(n-1),3*x(n-1)-2*x(n-2))
      d=Ø
      do 2 i=1,n
2     d=d+abs(x(i)-y(i))
      return
      end
```

```
      subroutine b3(x,y,n,d)
      dimension x(n),y(n)
      y(1)=amed3(x(1),x(2),3*x(2)-2*x(3))
      y(2)=amed3(x(1),x(2),x(3))
      do 1 i=3,n-2
      u=amed3(x(i-1),x(i),x(i+1))
      v=amed3(x(i-1),1.5*x(i-1)-Ø.5*x(i-2),x(i))
      w=amed3(x(i+1),1.5*x(i+1)-Ø.5*x(i+2),x(i))
1     y(i)=amed3(u,v,w)
      y(n-1)=amed3(x(n-2),x(n-1),x(n))
      y(n)=amed3(x(n),x(n-1),3*x(n-1)-2*x(n-2))
      d=Ø
      do 2 i=1,n
2     d=d+abs(x(i)-y(i))
      return
      end
```

```
      function amed3(u,v,w)
      amed3=v
      if(u.le.v.and.v.le.w) return
      if(u.ge.v.and.v.ge.w) return
      amed3=u
      if(v.le.u.and.u.le.w) return
      if(v.ge.u.and.u.ge.w) return
      amed3=w
      return
      end


      subroutine a3R(x,y,z,n,d)
      dimension x(n),y(n),z(n)
      call a3(x,y,n,d)
      if(d.eq.Ø) return
1     call a3(y,z,n,d)
      do 2 i=1,n
2     y(i)=z(i)
      if(d.gt.Ø) goto 1
      d=Ø
      do 3 i=1,n
3     d=d+abs(x(i)-y(i))
      return
      end


      subroutine b3R(x,y,z,n,d)
      dimension x(n),y(n),z(n)
      call b3(x,y,n,d)
      if(d.eq.Ø) return
1     call b3(y,z,n,d)
      do 2 i=1,n
2     y(i)=z(i)
      if(d.gt.Ø) goto 1
      d=Ø
      do 3 i=1,n
3     d=d+abs(x(i)-y(i))
      return
      end
```

```
      subroutine split(x,y,n,d)
      dimension x(n),y(n)
      do 1 i=1,n
1     y(i)=x(i)
      if(x(2).ne.x(3)) goto 2
      if(x(1).le.x(2).and.x(3).le.x(4)) goto 2
      if(x(1).ge.x(2).and.x(3).ge.x(4)) goto 2
      y(2)=x(1)
      y(3)=amed3(x(3),x(4),3*x(4)-2*x(5))
2     do 3 i=3,n-2
      if(x(i).ne.x(i+1)) goto 3
      if(x(i-1).le.x(i).and.x(i+1).le.x(i+2)) goto 3
      if(x(i-1).ge.x(i).and.x(i+1).ge.x(i+2)) goto 3
      y(i)=amed3(x(i),x(i-1),3*x(i-1)-2*x(i-2))
      y(i+1)=amed3(x(i+1),x(i+2),3*x(i+2)-2*x(i+3))
3     continue
      if(x(n-1).ne.x(n-2)) goto 4
      if(x(n).le.x(n-1).and.x(n-2).le.x(n-3)) goto 4
      if(x(n).ge.x(n-1).and.x(n-2).ge.x(n-3)) goto 4
      y(n-1)=x(n)
      y(n-2)=amed3(x(n-2),x(n-3),3*x(n-3)-2*x(n-4))
4     d=Ø
      do 5 i=1,n
5     d=d+abs(x(i)-y(i))
      return
      end


      subroutine a3RSR(x,y,z,w,n,d)
      dimension x(n),y(n),z(n),w(n)
      call a3R(x,y,z,n,d)
1     call split(y,z,n,d)
      call a3R(z,y,w,n,d)
      do 2 i=1,n
2     z(i)=x(i)-y(i)
      if(d.eq.Ø) return
      goto 1
      end


      subroutine b3RSR(x,y,z,w,n,d)
      dimension x(n),y(n),z(n),w(n)
      call b3R(x,y,z,n,d)
1     call split(y,z,n,d)
      call b3R(z,y,w,n,d)
      do 2 i=1,n
2     z(i)=x(i)-y(i)
      if(d.eq.Ø) return
      goto 1
      end
```

The following subroutine may be used to produce the map of the United States shown in Section 2 of Chapter 6.

```
        subroutine usmap(m)
        dimension m(5Ø)
        logical*1 d
        data d/'*'/
        write(6,1) m(2),(d,i=1,11),m(47),(d,i=1,22),
        +m(19),d,d,m(34),(d,i=1,8),m(26),m(23),(d,i=1,6),
        +m(45),m(29),d,m(37),m(49),(d,i=1,5),m(32),
        +m(21),d,d,m(12),m(41),m(22),d,d,m(7),d,m(39),d,
        +m(5Ø),(d,i=1,4),m(38),m(3Ø),d,d,m(27),m(15),
        +m(14),m(35),m(2Ø),d,m(8),d,m(28),m(13),m(48),d,
        +d,d,m(5),m(44),m(6),m(25),m(17),m(46),d,d,d,
        +m(16),m(33),d,d,d,m(42),d,d,d,m(3),m(36),m(4),
        +m(4Ø),(d,i=1,5),m(31),m(1),m(1Ø),(d,i=1,4),
        +m(24),(d,i=1,8),m(43),m(18),(d,i=1,31),m(9),d,
        +m(11),d,d,d
1       format(i3,1x,7a1,39x,3a1,x/3x,a1,i4,2x,21a1,18x,
        +a1,i3,a1/2x,a1,19x,i3,6x,a1,x,2a1,12x,3a1,2x,a1,
        +x/2x,a1,12x,i3,1Øx,i3,2a1,2x,2a1,x,a1,7x,a1,2i3,
        +x/x,a1,i4,26x,i3,x,2a1,x,a1,4x,2a1,2i3,a1,x/x,
        +a1,8x,i3,1Øx,i3,1Øx,i3,a1,3x,a1,3x,i3,a1,i3/a1,
        +15x,i3,2Øx,4a1,2i3,a1,4x/a1,22x,i3,3x,i3,4x,i3,
        +x,i3,3x,i3,a1,i3,x/a1,5x,i3,24x,i3,6x,i3,3x,2a1,
        +4x/x,a1,i3,6x,i3,3x,i3,1Øx,i3,6x,i3,3x,i3,2a1,
        +4x/x,a1,22x,i3,18x,i3,2a1,4x/2x,a1,33x,i3,9x,a1,
        +5x/3x,2a1,5x,i3,13x,i3,2x,i3,9x,i3,2a1,6x/5x,
        +3a1,8x,i3,18x,i3,i4,x,a1,8x/8x,3a1,23x,i3,7x,a1,
        +9x/11x,7a1,6x,i3,4x,i3,2x,7a1,2x,a1,8x/18x,2a1,
        +x,2a1,6x,8a1,6x,a1,2x,a1,7x/2Øx,a1,2x,a1,3x,2a1,
        +15x,a1,x,2a1,6x/24x,a1,x,a1,17x,i3,a1,6x/i3,22x,
        +2a1,19x,a1)
        return
        end
```

CHAPTER 7

FITTING

Give us the tools and we will finish the job

Sir Winston Churchill,
Radio Broadcast (1941).

INTRODUCTION

In this, the final chapter, we examine the general problem of fitting a statistical relation to a data array. We believe that this problem is a fundamental part of data analysis. Some would say that the topic of fitting is more appropriately dealt with in a book on statistical inference. In our opinion, this failure to make the distinction between fitting and inference has caused a lot of confusion among statisticians, and is one reason why analytic methods are not used as effectively in scientific work as they could be.

Why is fitting fundamental to exploratory data analysis? The answer can be seen by reviewing the material covered in the preceding six chapters. In each chapter our basic concern was to decompose the data into a systematic part (or smooth or structured part) and a residual part (or rough or unstructured part). In the first chapter this decomposition had very simple components: the data array consisted of a single batch of numbers, the systematic part was just a constant representing the typical value in the batch, and the residuals comprised the original batch, simply shifted by the subtraction of the typical value. In Chapter 2, where our concern was the comparison of several batches, we were performing a decomposition of the data using the relation

$$y(i,j) = a(i) + z(i,j)$$

where y(i,j) denotes the i-th number in the j-th batch, a(i) denotes the typical value for the j-th batch, and z(i,j) represents the residual from the i-th number in batch j. Continuing use of algebraic notation, we were concerned in Chapter 3 with fitting a straight line relation

$$y(i) = a + b*x(i) + z(i)$$

for a batch of pairs of numbers {x(i), y(i)}, where z(i) denotes the residual after the relation has been subtracted from the i-th response y(i). In Chapter 4, in which the problem of biological assay was introduced, the relations

$$s(i) = a1 + b*x(i) + z(i,1)$$

$$u(i) = a2 + b*x(i) + z(i,2)$$

were used to decompose the data in the parallel line assay, where s(i), u(i), and x(i) denote the standard response, unknown response and (logarithm of) dose level, respectively, for the i-th subject. The main discussion in Chapter 5 involved the fitting of row-plus-column relations to two-way tables, in other words, decomposing the data into the relation

$$y(i,j) = t + r(i) + c(j) + z(i,j)$$

where y(i,j) is the response or data array element in row i and column j, t is a typical value, r(i) a row effect and c(j) a column effect, and z(i,j) the residual in position (i,j). Then in Chapter 6 we were concerned with the sequence smoothing problem, or the decomposition

$$y(i) = s(i) + z(i)$$

where s(i) is the smooth and z(i) is the rough corresponding to the data response y(i) at position i in the sequence, and, in multiple regression, with the fitting of the equation

$$y(i) = a + b1*x1(i) + b2*x2(i) + \ldots + z(i)$$

to a batch of n-tuplets {x1(i), x2(i), ... , y(i)}.

Thus, if this book is about exploratory data analysis, fitting forms an integral part of such analysis. Of course there are other components of data analysis besides the fitting of relations, including methods of display, reexpressions, and even the choice of the relation to be fitted, but it is clear that the problem of fitting is a unifying thread. Accordingly it is natural in this final chapter to develop the theory of fitting a relation to data.

In the preceding chapters the fitting techniques used involved the use of medians. The median was suggested as an acceptable estimate for the typical value, both for summarizing single batches and comparing several batches. When straight line relations were fitted, the suggested procedure was based on taking medians of groups of x- and y-values. In the fitting of row-plus-column relations to two-way tables, the median polish procedure involves iterating until the row and column medians of the residuals are equal to zero. Three term running medians were used as the basis for effective sequence smoothers, and the procedure for performing a multiple regression involved repetition of the straight line fitting technique, itself based on medians.

While medians have served us well in our procedures for exploratory data analysis, it turns out that we can do better by imbedding the fitting procedure in a more general framework. Despite their simplicity when dealing with small data arrays when we use pencil and paper, we want something better than medians in general, for the following reasons:

(a) For large arrays computation of the median, whether by hand or by electronic computer, is time-consuming, since it usually involves sorting the numbers in the array. For a data array of size n, sorting requires n*log(n) operations, a quantity that increases much faster than n.

(b) The median estimate of typical value of a batch does not make use of the configuration, or shape, of the distribution of the numbers in the batch. In particular, outlying numbers have the same influence on the median as numbers near the center of the distribution.

(c) To develop improved fitting methods, we need a procedure that is sufficiently general to allow a variety of choices from which, through insight and experience, more effective techiques will emerge. The median is not easily generalized in this sense.

With these considerations in mind, in 1971 Frank Hampel suggested a general fitting procedure based on the use of what he called the influence function. The technique is documented in the monograph Robust Estimates of Location: Survey and Advances, coauthored by David Andrews, Peter Bickel, Frank Hampel, Peter Huber, William Rogers, and John Tukey, and published in 1972 by Princeton University Press. In the following sections we show how the concept of the influence function may be used to fit relations, for each of the situations considered in the first six chapters.

CENTERING

In this section the general problem of estimating the center of a distribution is considered. We assume that the batch of numbers has a reasonably symmetric distribution. (If not, it may be possible to make the distribution more symmetric by means of some transformation, as we discussed in Chapter 1.)

Imagine that the numbers in the batch are represented by particles, which are located in a one dimensional space like beads strung along a wire, their positions or coordinates being determined by their numerical values. Recall that such a representation was used in the construction of the box plot, in Chapter 1. Thus the smallest number in the batch would correspond to the left-most particle, and the largest number to the right-most particle.

Now consider the problem of choosing the position in the space that in some sense best represents the position of the center of the batch, and that could usefully be taken as a typical value for the numbers in the batch. One particular choice is the median, which could be obtained by starting at the left-most particle, and moving to the right, continually stepping across particles until half were on each side. (If the size of the batch is even, we decide, by symmetry, to locate the median halfway between the middle two numbers.) The median may thus be defined as the position in the space at which the sum of the influences of the particles is zero, where the "influence" of a particle on a point is +1 if the particle is to the left of the point, -1 if the particle is to its right.

In this definition, the "influence" of a particle on a point can only take the values +1 and -1, and it does not depend on how far away the particle is. More generally, we may incorporate some dependence on distance, and define the influence of a particle with coordinate x on a point with coordinate t as inf(x-t), where inf(.) is a general symmetric [that is, inf(-z)=-inf(z)] function. Considering the influences from all the particles in our representation of the batch, we may define the total influence on the point t as

$$\inf\{x(1)-t\}+\inf\{x(2)-t\}+\ldots+\inf\{x(n)-t\}$$

where x(i) is the value of the i-th number in the batch, assumed to contain n numbers in all.

The center of the batch can now be defined, quite simply, as the value of t for which the preceding sum is zero. This definition yields the median as the center in the special case when inf(z) = sign(z), that is, when inf(z) is +1 if z>0, -1 if z<0, and 0 if z=0. Another special case, important historically because much of the theory of statistical inference is based on it, is the mean, which arises when inf(z)=z.

Before proceeding to discuss the choice of the shape of the influence function, let us consider the computation of the center of a batch of numbers using the preceding definition. When inf(z)=z, the equation determining the center reduces to

$$\{x(1)-t\}+\{x(2)-t\}+\ldots+\{x(n)-t\} = 0$$

which may be solved explicitly for t to give

$$t = \{x(1)+x(2)+\ldots+x(n)\}/n$$

the mean of the batch. However for more general influence functions things are not quite so simple, and some kind of iteration is needed. To see how such an iteration would work, suppose we begin with some initial guess for the center, such as the median of the batch. At this initial point we can compute the total influence of the numbers in the batch. If this sum is positive, we need to move to the right, since the influence from the points to the right is outweighing the influence from the points to the left. Having moved a little way, we recompute the total influence, and again move by a small amount in the appropriate direction. Repeating this process, we will eventually arrive at a point at which the total influence is acceptably close to zero. By definition this point is the center of the batch.

The only problem with this approach is that

there does not seem to be a simple way of knowing how much to move at each step, and it may take a very long time to arrive at the desired goal. To derive a more effective iteration procedure, we need to do a little analysis. Define the function w(.) to be

$$w(z) = inf(z)/z$$

so that the influence function may be written as

$$inf(z) = z*w(z)$$

Let us now make the assumption that the function w(z) does not change appreciably as its argument changes. Some influence functions fit this requirement better than others: the influence function corresponding to the mean, inf(z)=z, fits it perfectly, for in that case w(z) = z/z = 1; for the median, however, we have inf(z) = sign(z), so it follows that w(z) = sign(z)/z = |1/z|, which is infinite at z=0 and approaches zero as z increases to infinity. With such notation the equation determining the center of a batch of n numbers x(1), x(2), ..., x(n) becomes

$$\{x(1)-t\}*w\{x(1)-t\} + \ldots + \{x(n)-t\}*w\{x(n)-t\} = 0$$

If the function w(.) were constant, we could solve this equation for t to obtain

$$t = \{x(1)*w(1)+\ldots+x(n)*w(n)\}/\{w(1)+\ldots+w(n)\}$$

where w(i) = w{x(i)-t}, for i=1,2,...,n. Since w(.) is not, in general, constant, this formula cannot be used, since the right-hand side of the formula involves the unknown value of t through the argument of w(i). However we can use the same kind of iteration suggested in the preceding paragraph with this formula, since, starting with some initial guess, we can obtain a new estimate for t from the formula, and recompute the right-hand side using the updated estimate, and thus obtain a further estimate.

Proceeding in this way, we obtain a sequence

of estimates of the center that converge (we hope) to the desired solution to the equation. The procedure may be stated algorithmically as follows:

(1) Put m=0 and choose some number, t(m), say, as an initial estimate of the center;

(2) For i=1,2,...,n compute the weights

w(i) = w{x(i)-t(m)}, and put

t(m+1)={x(1)*w(1)+..+x(n)*w(n)}/{w(1)+..+w(n)};

(3) If |t(m+1)-t(m)| is sufficiently small, end;

(4) Replace m by m+1, and go to (2);

This algorithm appears to converge for any starting point t(0), provided the influence function is reasonably well-behaved. However the precise conditions under which convergence is guaranteed are still not completely known.

Let us turn now to the choice of the influence function itself. Remember that we started with the idea of replacing the median with an estimate of center not entirely insensitive to the configuration of the batch. It is reasonable to insist that the influence function we choose should have two properties:

(a) the influence of faraway values is small

(b) the influence of closeby values is proportional to their distance

It is hard to argue against the desirability of the first property, for it provides a kind of resistance to or insulation against the effects of large outliers or blunders in the data - otherwise we would be allowing the tail to wag the dog, so to speak. The need for the second property is not quite so obvious. Expressed in terms of the weight function w(.) instead of the influence function inf(.), it requires that the weight function be constant when its argument is small. In the algorithm for iteratively computing the center, the estimate at each stage is a weighted average of the numbers in the batch, so it is clear that property (b) requires that for those numbers close to the center of the batch, the estimate of the center be obtained simply by taking their mean.

If we seek the simplest influence function that meets requirements (a) and (b), it is hard to improve on the polynomial

inf(z) = z*{1-(z/k)↑2}↑2, |z|<k

where inf(z) = Ø if |z| > k. This function increases between -k/2 and k/2, and decreases to zero outside this range. Thus the influences of the numbers in the batch increase with distance up to a point, but beyond a certain distance the influences get smaller. The weight function corresponding to this influence function is

w(z) = {1-(z/k)↑2}↑2, |z|<k

where w(z) = Ø if |z|>k. This particular form of weight function was suggested by John Tukey in lectures at Princeton University, and the estimator that results from its use is known as the biweight, the name arising from the fact that the double quadratic form of w(z) is sometimes called the bisquared function.

If we agree to use the biweight estimate of center, the problem of determining the scale constant k remains. This constant should depend on some measure of the size of the residuals, such as their midspread, and it is reasonable to put

k = c*s

where s is a measure of spread in the batch and c is a fixed constant. In Chapter 5 we discussed two measures of size of residuals after a row-plus-column relation had been fitted to a data array. One measure was the midspread of the distribution of the residuals: the other was the sum of the magnitudes of the residuals. Both measures have disadvantages - the first can be time-consuming to compute because some sorting is involved, while the second is influenced more than one would wish by large outliers. However, since no better simple measure is known, we have to be content with one or the other, and we shall use the mean of the magnitudes of the residuals as our measure of the scale s.

The complete algorithm for computing the

center of a batch of numbers may now be stated as

(1) Put m=0 and choose some initial estimate of center t(m) and spread s(m);

(2) For i=1,2,...,n compute the weights

w(i) = {1-minimum[1,u(i)]}↑2, where

u(i)=[{x(i)-t(m)}/{c*s(m)}]↑2, and then put

t(m+1)={x(1)*w(1)+..+x(n)*w(n)}/{w(1)+...+w(n)};

(3) If |t(m+1)-t(m)|/s(m) is sufficiently small, end

(4) Compute the mean absolute size of residuals,

s(m+1) = [|x(1)-t(m)|+...+|x(n)-t(m)|]/n;

(5) Replace m by m+1, and go to (2):

The only indeterminacy that remains is the choice of the constant c. This choice is to some extent a matter of philosophy, and to some extent depends on the nature of the data. The smaller the value of c, the more protection the estimator has against the influences of outliers. If there are no outliers, it may be best to choose a moderate to large value of c. Since c/2 is the number of multiples of the mean absolute size of residuals after which the influences begin to taper off, a value of c equal to 4 is probably the most protection one would ever want, while c=10 or more is not much different from using the mean as an estimate of center. Thus values of c between 4 and 10 are reasonable.

As an illustration of the technique of constructing the biweight estimator of center, we shall go through the computation for a specific example. The following eight numbers are the percentages of lobsters recaptured from tagged populations released from eight locations off Long Island between April 1968 and June 1969, and recaptured before April 1970. (Source - article by R.A. Cooper and J.R. Uzmann, Science, January 1971, pages 288 to 290.) The percentages are 13.0, 7.0, 9.4, 7.3, 6.4, 8.8, 6.2, and 5.5.

To calculate the biweight, let us use zero as an initial estimate of center and the mean of the absolute values (7.95) as an initial estimate of size of residuals. (We can certainly do better than this, by using the median as the initial estimate of center, for example, but zero requires less effort.) Let us also choose c=10 , and suppose we decide that the procedure has converged when the ratio |t(m+1)-t(m)|/s(m) is less than 0.01. The values of t(m), s(m) and w(1), w(2), ..., w(8) at each step of the iteration are

m	t(m)	s(m)	w(1)	w(2)	w(3)	w(4)	w(5)	w(6)	w(7)	w(8)
0	0.00	7.95	0.95	0.98	0.97	0.98	0.99	0.98	0.99	0.99
1	7.92	1.83	0.85	1.00	0.99	1.00	0.99	1.00	0.98	0.97
2	7.87	1.82	0.85	1.00	0.99	1.00	0.99	1.00	0.98	0.97
3	7.86									

Thus, after three steps, the estimate has converged to the value 7.86. Notice that the weights are all close to one, so that the biweight estimate is almost equal to the mean of the batch in this case. Consider now what happens if we choose c=4:

m	t(m)	s(m)	w(1)	w(2)	w(3)	w(4)	w(5)	w(6)	w(7)	w(8)
0	0.00	7.95	0.69	0.90	0.83	0.90	0.92	0.85	0.92	0.94
1	7.76	1.79	0.21	0.98	0.90	0.99	0.93	0.96	0.91	0.81
2	7.44	1.71	0.12	0.99	0.84	1.00	0.95	0.92	0.94	0.84
3	7.31	1.68	0.08	1.00	0.82	1.00	0.96	0.90	0.95	0.86
4	7.27	1.68	0.07	1.00	0.81	1.00	0.97	0.90	0.95	0.87
5	7.25									

In this case five steps were required for convergence, according to the criterion we are using. It is interesting to note that the weight associated with the first number in the batch (13.0) is progressively reduced to a value close to zero, so that this number is effectively eliminated from the calculations, while the weights associated with the other members of the batch are all close to one.

In fact, the more resistant biweight estimator of center is quite effective in this example. The numbers of tagged lobsters released from the various locations were 46, 975, 223, 521, 2412, 530, 857, and 146, respectively. One would expect that the errors in the recapture percentages would decrease as the released population increased, so that the first, third, and last numbers in the sample, particularly the first, would be the more errorprone. Looking at the weights associated with these numbers by the biweight procedure, we find w(1)=0.07, w(3)=0.81, and w(8)=0.87, which are the smallest three weights, all the others being 0.90 or more.

COMPARISONS

In this section we consider the problem of comparing the centers of several batches. We assume that each batch has approximately the same spread and a reasonably symmetric distribution. Recall from Chapter 2 that it is often possible to even out the spreads of the distributions of several batches by means of some reexpression of the data. While both evenness of spreads and symmetry are desirable when comparing batches, it may not be possible to accomplish both objectives with the same transformation. In such cases evening out the spreads is more important.

In the preceding section we assembled the machinery for effective estimation of the center of a single batch. When dealing with several batches, one might think that there is nothing more to be done - it is simply a matter of estimating the center separately for each batch. However to do this would be wasteful to some extent.

In the estimation technique described in the preceding section, it was necessary to estimate the size of residuals, so that the point at which the influence function began to return to zero could be determined. If we have several batches of data for each of which a center is to be estimated, it makes sense to use the same estimate of size of residuals for each batch, particularly since we are assuming that these spreads do not differ systematically from batch to batch. We may expect that the estimate of residual size obtained by combining the residual distributions from the various batches would be more accurate than that obtained from a single batch. This idea of combining estimates of residual size is called borrowing. Clearly the value of borrowing depends on the validity of the assumption of equal spreads in the various batches - if there are significant differences in these spreads, it may be better not to combine the residual distributions.

Let us now assemble a procedure for simultaneous centering of several batches. To keep the notation simple, suppose that the data array is composed of nr rows and nc columns, with the columns corresponding to the different batches. Thus for the moment we are assuming that the batches are each of size nr.

Using an algebraic notation, we denote the number in position (i,j) by y(i,j), and the estimate of the center of the j-th batch at the m-th iteration by a(j,m). When we extend the algorithm given in the preceding section to cope with several batches, the simultaneous estimation procedure based on the biweight proceeds as follows:

(1) Put $m=0$ and choose initial estimates a(1,m), a(2,m), ..., a(nc,m) for the center estimates and s(m) for the size of residuals;

(2) For $j=1,2,\ldots,nc$ and $i=1,2,\ldots,nr$, compute the weights

$$w(i,j) = \{1-\text{minimum}[1,u(i,j)]\}\uparrow 2, \quad \text{where}$$

$$u(i,j) = [\{y(i,j)-a(j,m)\}/\{c*s(m)\}]\uparrow 2,$$

and put a(j,m+1) equal to

$$\{y(1,j)*w(i,j)+..+y(nr,j)*w(nr,j)\}/W, \quad \text{where}$$

$$W = w(1,j)+...+w(nr,j);$$

(3) Put s(m+1) equal to the sum

$$\{|y(1,1)-a(1,m)|+...+|y(nr,nc)-a(nc,m)|\}/n$$

where n is the total number of elements in the array;

(4) If $|s(m+1)-s(m)|/s(m)$ is sufficiently small, end

(5) Replace m by m+1, and go to (2):

In comparing this algorithm to the earlier one, note that we have replaced the convergence criterion by one that involves simply the percentage change in the residual size s(m), rather than the center estimates a(j,m), j=1,2,...,nc. It is easy to generalize the algorithm to handle different numbers of observations in the different batches.

To illustrate the application of the algorithm, we return to the data array "bees" introduced in Chapter 4. This array consists of the amounts of solution consumed by honeybees at various sulfur concentrations, and has eight batches, each containing eight numbers. It is listed as follows:

Batch 1	2	3	4	5	6	7	8
12	8	84	57	43	87	80	130
4	6	29	28	95	114	72	72
5	8	13	39	114	44	60	127
5	14	9	36	39	57	77	69
4	4	17	22	51	20	92	86
2	7	16	27	55	90	24	81
3	10	15	20	47	69	72	81
2	4	19	51	61	71	71	76

Let us compute the biweight estimators, using a common scale estimate. As in the preceding example we shall take, for simplicity, the initial estimates to be all zero, and the initial estimate of s as the mean absolute value of the numbers in the data array. With c=10 we get the following results:

m	s(m)	a(1)	a(2)	a(3)	a(4)	a(5)	a(6)	a(7)	a(8)
0	45.4	0.0	0.0	0.0	0.0	0.0	0.0	0.0	0.0
1	13.1	4.6	7.6	24.8	34.9	62.2	68.0	68.1	89.2
2	12.7	4.6	7.6	22.5	34.9	61.0	69.2	69.5	88.8
3	12.7	4.6	7.6	22.1	34.9	60.7	69.5	69.7	88.6

Thus the procedure converges after three iterations. The weights w(i,j) when m=2 are tabulated as follows:

1.0	1.0	0.6	0.9	1.0	1.0	1.0	0.8
1.0	1.0	1.0	1.0	0.9	0.8	1.0	1.0
1.0	1.0	1.0	1.0	0.7	0.9	1.0	0.8
1.0	1.0	1.0	1.0	0.9	1.0	1.0	1.0
1.0	1.0	1.0	1.0	1.0	0.7	0.9	1.0
1.0	1.0	1.0	1.0	1.0	1.0	0.8	1.0
1.0	1.0	1.0	1.0	1.0	1.0	1.0	1.0
1.0	1.0	1.0	1.0	1.0	1.0	1.0	1.0

Inspecting these weights, we see that all are greater than 0.6, and most are close to 1. The result of the fitting is thus not vastly different from what we would get by taking the simple mean of each column. Let us now see what happens when we put c=4:

m	s(m)	a(1)	a(2)	a(3)	a(4)	a(5)	a(6)	a(7)	a(8)
0	45.4	0.0	0.0	0.0	0.0	0.0	0.0	0.0	0.0
1	12.8	4.6	7.6	22.3	34.2	57.0	62.1	65.6	82.8
2	12.0	4.6	7.6	17.0	34.0	51.5	66.2	72.6	78.5
3	12.0	4.6	7.6	16.7	33.8	49.7	68.8	74.3	77.6

The procedure again converges after three steps, but the estimated centers are actually quite different in the two cases. There is little doubt that the fitting with c=4 gives the superior result, since the typical values increase much more smoothly in this case. The more resistant procedure is quite effective in downweighting some of the observed values, as can be seen from the following table of values of w(i,j) when m=2:

1.0	1.0	0.0	0.6	0.9	0.7	1.0	0.0
1.0	1.0	0.9	1.0	0.0	0.0	1.0	1.0
1.0	1.0	1.0	1.0	0.0	0.6	0.9	0.0
1.0	1.0	0.9	1.0	0.9	0.9	1.0	0.9
1.0	1.0	1.0	0.9	1.0	0.0	0.7	1.0
1.0	1.0	1.0	1.0	1.0	0.6	0.0	1.0
1.0	1.0	1.0	0.8	1.0	1.0	1.0	1.0
1.0	1.0	1.0	0.8	0.9	1.0	1.0	1.0

Of course in this example we have violated our rule to some extent, because the spreads of the data are not the same in each batch. It would have been better if some transformation had been applied to the data, to even out the spreads, before carrying out the fitting procedure.

RELATIONS

Having discussed the centering problem in the preceding two sections, we turn now to the problem of fitting a general relation to a data array. To cope with this more general fitting problem, we need to introduce some more machinery.

When finding a typical value for one or more batches we defined the center as the point at which the total influence of the data is equal to zero. For a single batch, this led to an estimate satisfying the formula

t = {x(1)*w(1)+...+x(n)*w(n)}/{w(1)+...+w(n)}

where x(1), x(2), ..., x(n) are the numbers in the batch, and w(1), ..., w(n) are iteratively calculated weights.

Consider the problem of fitting a straight line y(i)=a+b*x(i) to a batch of pairs {x(i),y(i)}. In this case there are two parameters to be determined, a and b. By analogy with the centering problem, it is natural to require that the sum of the influences of the residuals be zero, that is

inf{y(1)-a-b*x(1)}+...+inf{y(n)-a-b*x(n)} = 0

Expressing this in terms of the weights, we have

{y(1)-a-b*x(1)}*w(1)+...+{y(n)-a-b*x(n)}*w(n) = 0

where the values of w are obtained iteratively. The problem is that this equation can determine only one of the unknown parameters a and b. To determine both parameters, we need another equation.

To get a second equation, we can make use of another requirement, one that is fundamental to data analysis, and which we have used repeatedly. This requirement is that the residuals should be without structure. In preceding chapters we have used the idea of lack of structure or pattern intuitively, without defining it properly. It is now time to be more specific. When we fit relations to data, our objective is to decompose the data into a smooth part and a residual part, where the smooth part has as much structure as possible and the residual as little structure as possible. It follows that there can be no relation between the smooth part and the residuals, for otherwise it would be possible to remove this relation and thus get more structure into the smooth part and less into the residuals. What we are really saying, then, when we say that residuals should be unstructured is that there should be no relation between the residuals and the fitted part.

How can we define the concept "no relation"? A very simple definition is to say that two numbers have no relation if one of them is zero. If we think of the two numbers as coordinates in an (x,y) plot, then points close to either of the two axes have little in common, while those close to the line whose equation is y=x have a lot in common. (If the numbers are equal in magnitude but opposite in sign, they have a strong negative relation, which is not the same thing as having zero relation.) Turning now to a batch of pairs of numbers {(x(i),y(i)}, i=1,2,...,n, we may say that the components have no relation if

$$x(1)*y(1)+x(2)*y(2)+...+x(n)*y(n) = 0$$

In this equation each point {x(i),y(i)} is counted with equal weight in the summation. If we say that there is to be no relation between a residual and a smooth part in a decomposition of some data, we will probably want to attach more weight to some residuals than to others, depending on their size. The natural weights to use are, of course, the values of w(i) that arise in the centering problem. Since the i-th residual is y(i)-a-b*x(i) and the i-th smooth part is a+b*x(i), we accordingly arrive at the requirement that

$$\{a+b*x(1)\}*\{y(1)-a-b*x(1)\}*w(1) + ... + \{a+b*x(n)\}*\{y(n)-a-b*x(n)\}*w(n) = 0$$

We now have two equations which may be solved simultaneously for a and b. These equations are equivalent to the pair of equations

$$\{y(1)-a-b*x(1)\}*w(1)+...+\{y(n)-a-b*x(n)\}*w(n) = 0$$

$$x(1)*\{y(1)-a-b*x(1)\}*w(1)+..+x(n)*\{y(n)-a-b*x(n)\}*w(n)=0$$

which have a very simple interpretation. The smooth

part, a+b*x(i), consists of two terms, a constant term, whose coefficient or parameter is a, and a carrier, x(i), whose coefficient is b. The first of these equations states that there is no relation between the residuals and the constant component, while the second equation states that there is no relation between the residuals and the carrier.

Let us now consider how we would go about solving the two equations for a and b. First define

$$xt = \{x(1)*w(1)+...+x(n)*w(n)\}/\{w(1)+...+w(n)\}$$

$$yt = \{y(1)*w(1)+...+y(n)*w(n)\}/\{w(1)+...+w(n)\}$$

Thus xt and yt are weighted averages of the x- and y-values, using the weights w. In a sense they are estimates of the centers of the distributions of the x-values and the y-values, respectively. Now define, for i=1,2,...,n,

$$\tilde{x}(i) = x(i)-xt, \quad \tilde{y}(i) = y(i)-yt$$

Substituting $x(i)=\tilde{x}(i)+xt$ and $y(i)=\tilde{y}(i)+yt$ and using the definitions of xt and yt, the equations determining a and b reduce to

$$\{a-yt+b*xt\}*\{w(1)+w(2)+...+w(n)\} = 0$$

$$\tilde{x}(1)*\{\tilde{y}(1)-b*\tilde{x}(1)\}*w(1)+...+\tilde{x}(n)*\{\tilde{y}(n)-b*\tilde{x}(n)\}*w(n) = 0$$

From these equations it follows immediately that

$$a = yt - b*xt$$

where b is given by the ratio

$$\frac{\tilde{x}(1)*\tilde{y}(1)*w(1)+...+\tilde{x}(n)*\tilde{y}(n)*w(n)}{w(1)*\tilde{x}(1)\uparrow 2+...+w(n)*\tilde{x}(n)\uparrow 2}$$

Thus if the weights are known, the parameters a and b are given explicitly by these formulas. However the weights are determined by the residuals, which can be computed only if a and b are known, so an iterative procedure is required. A reasonable algorithm would be described as follows:

(1) Put m=Ø and choose initial estimates for the parameters a(m) and b(m) and for the scale s(m). For i=1,2,...,n, compute the residuals

z(i) = y(i)-a(m)-b(m)*x(i);

(2) For i=1,2,...,n compute the weights

w(i) = {1-minimum[1,u(i)]}↑2, where

u(i) = [z(i)/{c*s(m)}]↑2, and put

xt = {x(1)*w(1)+..+x(n)*w(n)}/{w(1)+..+w(n)},

yt = {y(1)*w(1)+..+y(n)*w(n)}/{w(1)+..+w(n)};

(3) For i=1,2,...,n, replace x(i) by x(i)-xt and y(i) by y(i)-yt;

(4) For i=1,2,..,n put z(i)=y(i)-b(m+1)*x(i), where

b(m+1)={x(1)*y(1)*w(1)+..+x(n)*y(n)*w(n)}/S,

and S=w(1)*x(1)↑2+...+w(n)*x(n)↑2, and replace

y(i) by z(i). Put a(m+1) = yt-b(m+1)*xt;

(5) Put s(m+1) = {|z(1)|+|z(2)|+...+|z(n)|}/n;

(6) If |s(m+1)-s(m)|/s(m) is sufficiently small, end

(7) Replace m by m+1, and go to (2):

As an illustration of this algorithm, let us return to the acid density data, first considered in Chapter 3. Choosing the extremely naive initial estimates a(Ø)=Ø and b(Ø)=Ø, and taking s(Ø) to be the mean of the absolute values of the y-values, we obtain the following results for c=1Ø and a convergence criterion of Ø.Ø1:

m	s(m)	a(m)	b(m)	w(1)	w(2)	w(3)	w(4)	w(5)	w(6)
0	0.458	0.000	0.000	0.99	1.00	1.00	1.00	1.00	0.99
1	0.006	0.005	0.876	0.98	1.00	1.00	0.97	0.97	0.93
2	0.006	0.005	0.877						

The procedure thus converges after two steps. It is of interest to compare the values of the parameters a and b with those obtained in Chapter 3 by the median-based procedure. There the y-intercept, a, was 0.0065 and the slope, b, was calculated as 0.8775, so the values obtained from the iterative bi-weight procedure are in the same ballpark. However if we use the value c=4, we obtain a different result:

m	s(m)	a(m)	b(m)	w(1)	w(2)	w(3)	w(4)	w(5)	w(6)
0	0.458	0.000	0.000	0.93	0.98	1.00	0.99	0.98	0.93
1	0.006	0.005	0.877	0.83	1.00	0.99	0.87	0.85	0.54
2	0.006	0.004	0.881	0.86	1.00	0.99	0.90	0.90	0.34
3	0.006	0.002	0.886	0.89	1.00	1.00	0.94	0.96	0.10
4	0.005	-0.001	0.894	0.96	0.99	1.00	0.98	1.00	0.00
5	0.005	-0.003	0.900	0.99	0.98	1.00	0.99	1.00	0.00
6	0.005	-0.003	0.900						

In this case the point corresponding to the largest value of x has been given zero weight. It may be recalled that in our discussion of this example this point was observed to be out of line, so it is not surprising that the resistant fitting procedure ignored it. The line actually fits better using the resistant fitting method which ignores the outlier, as can be seen by examining the scatter plot of residuals and comparing it with the plot of residuals shown in Chapter 3.

This line-fitting procedure is easily generalized to allow for a situation in which there is more than one carrier, such as we discussed in the second section of Chapter 6. In this case the relation to be fitted may be expressed in algebraic terms as

y(i) = a*x0(i) + b1*x1(i) + b2*x2(i) + ... + bp*xp(i)

where there are p+1 carriers x0, x1, ..., xp, including the constant carrier x0(i)=1, for all i. There are p+1 parameters a, b1, b2, ..., bp to be found, so p+1 equations are needed to determine them. Using the requirement that the residuals should bear no relation to any of the carriers (including the constant term), we arrive naturally at the set of equations z(1)*w(1)+z(2)*w(2)+...+z(n)*w(n) = Ø

z(1)*x1(1)*w(1)+z(2)*x1(2)*w(2)+...+z(n)*x1(n)*w(n) = Ø

.. ..

z(1)*xp(1)*w(1)+z(2)*xp(2)*w(2)+...+z(n)*xp(n)*w(n) = Ø

where z(i) = y(i)-a-b1*x1(i)-...-bp*xp(i), and w(i) are iteratively determined weights, for i=1,2,...,n.

It is only slightly more difficult to solve these general equations than those the case p=1. The fitting can be carried out by a similar procedure to the one used for multiple regression in Chapter 6, that is, to sweep out the carriers in turn. In the present situation the method may be described algorithmically as follows:

(1) Put m=Ø and choose initial values a(m), b1(m), ..., bp(m) for the parameters and s(m) for the residual scale, and for i=1,2,...,n compute the residuals

z(i)=y(i)-a(m)-b1(m)*x1(i)-...-bp(m)*xp(i);

(2) For i=1,2,...,n compute the weights

w(i) = {1-minimum[1,u(i)]}↑2,

where u(i) = [z(i)/{c*s(m)}]↑2;

(3) For j=Ø,1,2,...,p, proceed as follows:-

For k=j+1,j+2,...,p, replace xk(i) by

xk(i)-b*xj(i), for i=1,2,...,n, where

b={xk(1)*xj(1)*w(1)+...+xk(n)*xj(n)*w(n)}/S, and

S = w(1)*xj(1)↑2+...+w(n)*xj(n)↑2;

Replace y(i) by y(i)-b*xj(i), for

i=1,2,...,n,

where b={y(1)*xj(1)*w(1)+...+y(n)*xj(n)*w(n)}/S;

(4) For i=1,2,...,n, put z(i) = y(i);

(5) Put s(m+1) = {|z(1)|+|z(2)|+...+|z(n)|}/n;

(6) If |s(m)-s(m+1)|/s(m) is sufficiently small, end

(7) Replace m by m+1, and go to (2):

In this algorithm the residuals and the fitted regression are obtained by a gradual process of sweeping out the effects of the carriers from each other and from the response, in turn. However the parameters a, b1,..., bp are not computed explicitly. To obtain these coefficients, it is necessary to do some kind of algebraic manipulation similar to that performed on the crime data in Chapter 6. Fortunately there is an easy way of incorporating the necessary manipulations into the algorithm. This procedure involves augmenting the carriers x0, x1, ..., xp and the response y each by p+1 numbers, and carrying through the arithmetic in the algorithm with n+p+1 numbers each time instead of n. The first carrier, x0, is augmented by the sequence (1,0,0,...,0), the second, x1, by (0,1,0,...,0), x2 by (0,0,1,...,0), etc. The numbers attached to the final carrier xp are (0,0,0,...,1), while p+1 zeros (0,0,0,...,0) are attached to y.

After the algorithm has converged, it may be shown that what was originally the y column (augmented by the p+1 zeros) has been replaced by the residuals in the first n positions and the fitted parameters a, b1, ..., bp in the last p+1 positions, with their signs reversed. To see why this is so, it is necessary to work through a little algebra, and this task is left to readers who are so inclined. It is fairly easy to see why the procedure works in the case p=1, when just two numbers are added to the columns of the data array.

As for the biweight centering algorithm for one or more batches, a computer subroutine for carrying out the multiple regression algorithm is given at the end of this chapter. The procedure is known as the <u>modified</u> <u>Gram-Schmidt</u> algorithm, and is generally regarded as the best procedure available, in terms of speed and numerical stability, for fitting a regression relation to a point cloud. One advantage, not shared by the median-based procedure given in Chapter 6, is that the order in which the carriers are specified does not affect the result.

As an illustration of the Gram-Schmidt procedure, let us return to the crime data originally presented in Chapter 6. The following numbers are the values of the regression coefficients and the residual absolute size at each iteration, starting with zero values for the coefficients and using c=8. The carriers, in order, are murder, assault, and percent urban.

m	s(m)	a(m)	b1(m)	b2(m)	b3(m)
0	1.285	0.00	0.00	0.00	0.0000
1	0.090	0.39	0.32	0.19	0.0036
2	0.089	0.35	0.29	0.19	0.0041
3	0.089	0.34	0.28	0.20	0.0042

The values obtained are not very different from those obtained earlier. In Chapter 6 we calculated the regression coefficients as a=0.41, b1=0.34, b2=0.20, and b3=0.0026.

TWO-WAY TABLES

In Chapter 5 we discussed the fitting of various relations to two-way tables, the most basic being the row-plus-column relation, which may be expressed algebraically as

$$x(i,j) = t + r(i) + c(j)$$

where x(i,j) is the number in row i and column j, t is a typical value, r(i) is a row effect for row i, and c(j) is a column effect for column j. The procedure of median polish was used to fit the relation. This procedure involves "sweeping out" row medians and column medians from the rows and columns of the data array, and repeating until the row and column medians of the residuals are all zero. In this section we propose to incorporate the fitting of the row-plus-column model into the general biweight technology.

The problem of fitting a row-plus-column relation to a two-way table is a straightforward generalization of the problem of simultaneous centering of several batches, considered earlier. In this case not only do we require that the centers of the distributions of residuals for each column be zero, we also want the centers of the residual distributions to be zero for each row. Given an array of weights w(i,j) computed from the residuals using the biweight function, these requirements may be stated algebraically as

z(1,j)*w(1,j)+...+z(nr,j)*w(nr,j) = Ø, j=1,2,...,nc

z(i,1)*w(i,1)+...+z(i,nc)*w(i,nc) = Ø, i=1,2,...,nr

where z(i,j) = x(i,j)-t-r(i)-c(j), and nr is the number of rows and nc the number of columns in the data array.

This set of equations may be solved to determine the parameters t, r(1),...,r(nr), and c(1),...,c(nc), provided that the w(i,j) are known. The w(i,j) can be determined from the residuals once the parameters are known, so it is clear how an iterative procedure would work. There is one problem, however, and that is that this system of nr+nc equations contains only nr+nc-1 independent equations. (This can be seen by noting that the sum of the first set of nc equations is the same equation as the sum of the second set of nr equations.) This could hardly be otherwise, of course, since there are only nr+nc-1 independent parameters. If a constant is subtracted from all the row effects, another constant subtracted from all the column effects, and the sum of the two constants added to the typical value, the fitted relation is unchanged.

To avoid the indeterminacy in the parameters, we need to impose two constraints. The constraints we shall choose are that the sum of the row effects and the sum of the column effects should both be zero, that is

r(1)+r(2)+...+r(nr) = Ø

c(1)+c(2)+...+c(nc) = Ø

We can now construct an algorithm for fitting the row-plus-column relation by using biweights as follows:

(1) Put m=Ø and choose initial estimates t(m), r(1,m),...,r(nr,m) and c(1,m),...,c(nc,m) for the parameters and s(m) for the size of residuals. Compute residuals z(i,j), where

z(i,j) = x(i,j)-t(m)-r(i,m)-c(j,m);

(2) For each pair (i,j), compute the weights

w(i,j) = {1-minimum[1,u(i,j)]}↑2, where

u(i,j) = [z(i,j)/{c*s(m)}]↑2, and form the sums

cn(j) = z(1,j)*w(1,j)+...+z(nr,j)*w(nr,j),

cd(j) = w(1,j)+...+w(nr,j), for j=1,2,...,nc,

and rn(i) = z(i,1)*w(i,1)+...+z(i,nc)*w(i,nc),

rd(i) = w(i,1)+...+w(i,nc), for i=1,2,...,nr;

(3) Put r(i,m+1)=r(i,m)+rn(i)/rd(i),

for i=1,2,...,nr, and, for j=1,2,...,nc,

c(j,m+1) = c(j,m)+cn(j)/cd(j), and

t(m+1)=t(m)-{cn(1)+..+cn(nc)}/{cd(1)+..+cd(nc)};

(4) Compute rs={r(1,m+1)+..+r(nr,m+1)}/nr,

and cs={c(1,m+1)+..+c(nc,m+1)}/nc,

and replace r(i,m+1) by r(i,m+1)-rs, for

i=1,2,...,nr, c(j,m+1) by c(j,m+1)-cs, for

j=1,2,...,nc, and t(m) by t(m+1)+rs+cs;

(5) For each (i,j) replace the residuals by

z(i,j)=x(i,j)-t(m+1)-r(i,m+1)-c(j,m+1). Put

s(m+1) = {|z(1,1)|+...+|z(nr*nc)|}/(nr*nc);

(6) If |s(m+1)-s(m)|/s(m) is sufficiently small, end

(7) Replace m by m+1 and go to (2):

The data array "deaths," considered in Chapter 5, will be used to illustrate the use of this algorithm. Starting with the very crude initial parameter estimates t(Ø)=Ø, r(i)=Ø for all values of i and c(j)=Ø for all j, and s(Ø) equal to the average magnitude of the numbers in the array, the following decomposition is obtained with c=1Ø. The residuals are given in the body of the table with the weights in parentheses, and the row effects are listed in the right-most column, with the column effects and typical value in the last row.

-1.1(1.Ø)	3.2(1.Ø)	-5.Ø(Ø.9)	2.7(1.Ø)	-19.7
-Ø.6(1.Ø)	Ø.3(1.Ø)	-2.Ø(1.Ø)	2.Ø(1.Ø)	-14.Ø
-Ø.7(1.Ø)	Ø.Ø(1.Ø)	1.8(1.Ø)	-1.2(1.Ø)	-5.1
-1.Ø(1.Ø)	-3.8(Ø.9)	5.Ø(Ø.9)	Ø.2(1.Ø)	9.4
4.1(Ø.9)	-Ø.3(1.Ø)	1.6(1.Ø)	-4.8(Ø.9)	29.5
1.8	-5.7	9.6	-5.6	3Ø.9

When the value of c is reduced to 4, three steps are needed before convergence is reached (as in preceding programs in this chapter, the convergence criterion is a reduction in s by less than 1 percent). The results for c=4 are as follows:

-1.Ø(Ø.9)	2.5(Ø.8)	-4.6(Ø.5)	1.9(Ø.9)	-19.3
-Ø.3(1.Ø)	-Ø.1(1.Ø)	-1.3(Ø.9)	1.4(Ø.9)	-13.9
-Ø.3(1.Ø)	-Ø.3(1.Ø)	2.6(Ø.8)	-1.6(Ø.9)	-5.Ø
Ø.6(1.Ø)	-2.9(Ø.8)	6.9(Ø.1)	Ø.9(1.Ø)	8.2
4.6(Ø.5)	-Ø.5(1.Ø)	2.4(Ø.8)	-5.2(Ø.4)	29.9
1.5	-5.6	8.9	-4.9	3Ø.8

The final value of s is found to be 2.Ø1 in each case. The decomposition is not affected greatly by the degree of resistance in the fitting procedure, since the results are much the same for the two values of c, and both are quite similar to the result obtained in Chapter 5 using median polish, allowing for the fact that the row and column effects have been constrained differently in the biweight fitting procedure.

The fitting of higher-order relations, such as the column-plus-row-times-column relation

$$y(i,j) = t + c1(j) + r(i)*c2(j)$$

is a straightforward generalization of the methods already developed. In this case the problem may be tackled by combining the fitting techniques for regression and two-way tables. By analogy with two-way table fitting we require that the set of equations

$$z(1,j)*w(1,j)+...+z(nr,j)*w(nr,j) = 0, \text{ for } j=1,2,...,nc$$

should be satified, where z(i,j) is the residual y(i,j)-t-c1(j)-r(i)*c2(j). These are the equations that arise for the column effects c(j) in the problem of fitting a row-plus-column relation to a data array. We also require that the residuals should bear no relation to the fitted part. Now if the values of c2(j) were known for j=1,2,...,nc, these could be regarded as comprising a carrier term in a regression, and the unrelatedness of the residuals with this carrier would give rise to the equation

$$z(i,1)*c2(1)*w(i,1)+...+z(i,nc)*c2(nc)*w(i,nc) = 0$$

which is then true for all values of i. Similarly, if the values of r(i) were known for i=1,2,...,nr, we would have a regression situation with r(i) as the carrier, so we would require

$$z(1,j)*r(1)*w(1,j)+...+z(nr,j)*r(nr)*w(nr,j) = 0$$

for each value of j, yielding a set of nc equations. Thus we have nc+nr+nc equations in all for the parameters. These equations may again be solved iteratively as before.

We have covered the basic ideas underlying the fitting of general relations to data. Of course this has been a very brief treatment, and a lot of concepts have been skimmed over rather quickly. As stated earlier, the iterative procedures we have suggested do not always converge, and even when they do, there is no guarantee that the solution obtained is a correct one, since the system of equations determining the parameters may not have a unique solution. Though some improvements can be made by using better guesses for the starting points for the iterations, numerical and statistical problems remain, and space does not permit us to give an adequate treatment here.

PROGRAM COMPONENTS

This section contains three routines for performing fitting procedures introduced in the text. The first, "center," computes the biweight estimators of the centers of several batches, each having the same size. This program has two parameters, "epsilon," the proportion by which the sum of absolute residuals must be reduced for iterations to continue, and "cutoff," the value of the constant (c) which occurs in the biweight estimator of center. The program "regress," which fits a multiple regression using the modified Gram-Schmidt procedure, has parameters "epsilon" and "cutoff," as before, and a third parameter, "atom," that is used to prevent the estimates of the coefficients from "blowing up" when one of the carriers is a linear combination of previously swept out carriers. A value of 0.0001 or smaller for "atom" is reasonable in most applications. (In the APL function "REGRESS" there is an additional parameter, "CONSTANT," that causes the constant carrier to be included when set to 1 and neglected when set to 0. The third program, "addfit," can be used to fit a row-plus-column relation to a two-way table, and has the same parameters as "regress," where in this case "atom" is used to avoid the possibility of a zero dision in case a whole row or column is assigned zero weight.

APL FUNCTIONS

In the APL functions which follow, residuals are returned in the array "Z" while fitted parameters are stored in the arrays "CE" (center estimates of columns in "CENTER"), "CF" (coefficients in "REGRESS") and "TV," "RE," and "CE" (typical value, row effects, and column effects in "ADDFIT").

```
      ∇CENTER[⎕]∇
      ∇ Z←CENTER X;N;R;W;S;SNEW
[1]     S←+/|,Z÷N←ρ,Z←X-+/R←(1↑ρX)ρ0
[2]     CE←(+⌿X×W)÷+⌿W←W×W←1-(1⌊Z÷CUTOFF×S)*2
[3]     ⎕←SNEW←(+/|,Z←X-R∘.+CE)÷N
[4]     →2×EPSILON<|1-(S←SNEW)÷S
      ∇


      ∇REGRESS[⎕]∇
      ∇ Z←REGRESS X;NR;NC;I;S;SNEW;Y;W;J;D;B;JJ
[1]     →3×⍳CONSTANT=0×NR←1↑ρX
[2]     X←(NRρ1),X
[3]     Z←(NC,NC+1)ρ(I∘.=I←⍳NC),(NC←¯1+1↓ρX)ρ0
[4]     S←+/(÷NR)×|Y←,X[;NC+1]
[5]     W←W×W←1-(1⌊Y÷CUTOFF×S)*2
[6]     J←1
[7]     D←NCρ0
[8]     →12×⍳ATOM≥D[J]←+/W×X[;J]*2
[9]     B←(X[;J]×W)+.×X[;JJ←J+⍳NC+1-J]÷D[J]
[10]    X[;JJ]←X[;JJ]-X[;J]∘.×B
[11]    Z[;JJ]←Z[;JJ]-Z[;J]∘.×B
[12]    →8×⍳NC≥J←J+1
[13]    ⎕←SNEW←(÷NR)×+/|Y←,X[;NC+1]
[14]    →5×⍳EPSILON<|1-(S←SNEW)÷S
[15]    CF←-Z[;NC+1]
[16]    Z←X[;NC+1]
      ∇


      ∇ADDFIT[⎕]∇
      ∇ Z←ADDFIT X;NR;NC;N;S;SNEW;W;CN;CD;RA;CA
[1]     RE←(NR←1↑ρX)ρ0
[2]     CE←(NC←1↓ρX)ρ0
[3]     S←(÷N←NR×NC)×+/|,X
[4]     Z←X-RE∘.+CE+TV←0
[5]     RE←RE+(+/W×Z)÷ATOM⌈+/W←W×W←1-(1⌊Z÷CUTOFF×S)*2
[6]     CE←CE+(CN←+⌿Z×W)÷CD←ATOM⌈+⌿W
[7]     RE←RE-RA←+/RE÷NR
[8]     CE←CE-CA←+/CE÷NC
[9]     TV←TV+RA+CA-(+/CN)÷+/CD
[10]    ⎕←SNEW←(÷N)×+/|,Z←X-TV+RE∘.+CE
[11]    →5×EPSILON<|1-(S←SNEW)÷S
      ∇
```

FORTRAN SUBROUTINES

```
        subroutine center(x,ce,n,nr,nc,cutoff,epsilon)
        dimension x(nr,nc),ce(nc)
        sar=0
        do 1 i=1,nr
        do 1 j=1,nc
        ce(j)=0
1       sar=sar+abs(x(i,j))
        s=sar/n
2       do 4 j=1,nc
        sn=0
        sd=0
        do 3 i=1,nr
        u=amin1(1.,((x(i,j)-ce(j))/(cutoff*s))**2)
        w=(1-u)**2
        sn=sn+w*x(i,j)
3       sd=sd+w
4       ce(j)=sn/sd
        sar=0
        do 5 j=1,nc
        do 5 i=1,nr
5       sar=sar+abs(x(i,j)-ce(j))
        if(abs(1-s*n/sar).le.epsilon) return
        s=sar/n
        goto 2
        end
```

In the following regression subroutine the input array, x, contains the carriers as columns and the response as the last column. If, as is usually the case, the regression is to contain a constant term, the first column of x should contain ones. The array z is required for temporary storage of the array of zeros and ones attached underneath the x array, as described in the text. The arrays w and d are also needed for temporary storage of weights and weighted sums of squares of carriers. Here nr is the number of rows, nc the number of carriers (including the constant), and nc1 is nc+1. When the procedure converges, the residuals are stored in the last column of x and the regression coefficients are stored in the last column of z, with their signs reversed.

```
        subroutine regress(x,z,w,d,nr,nc,nc1,cutoff,
       +epsilon,atom)
        dimension x(nr,nc1),z(nc,nc1),w(nr),d(nc)
        do 2 i=1,nc
        do 1 j=1,nc1
1       z(i,j)=0
```

```
2       z(i,i)=1
        sar=Ø
        do 3 i=1,nr
3       sar=sar+abs(x(i,ncl))
        s=sar/nr
4       do 5 i=1,nr
        u=aminl(1.Ø,(x(i,ncl)/(cutoff*s))**2)
5       w(i)=(1-u)**2
        do 1Ø j=1,nc
        d(j)=Ø
        do 6 i=1,nr
6       d(j)=d(j)+w(i)*x(i,j)**2
        if(d(j).lt.atom) goto 1Ø
        do 1Ø k=j+1,nc+1
        sxy=Ø
        do 7 i=1,nr
7       sxy=sxy+w(i)*x(i,j)*x(i,k)
        b=sxy/d(j)
        do 8 i=1,nr
8       x(i,k)=x(i,k)-b*x(i,j)
        do 9 i=1,nc
9       z(i,k)=z(i,k)-b*z(i,j)
1Ø      continue
        sar=Ø
        do 11 i=1,nr
11      sar=sar+abs(x(i,ncl))
        if(abs(1-s*nr/sar).lt.epsilon) return
        s=sar/nr
        goto 4
        end
```

In the following subroutine for fitting the row-plus-column relation, row effects are stored in an array re of size nr, column effects are kept in an array ce of dimension nc, while the typical value is returned in the parameter t.

```
        subroutine addfit(x,t,re,ce,n,nr,nc,cutoff,
        +epsilon,atom)
        dimension x(nr,nc),re(nr),ce(nc)
        dimension rn(2ØØ),rd(2ØØ),cn(5Ø),cd(5Ø)
        do 1 i=1,nr
1       re(i)=Ø
        do 2 j=1,nc
2       ce(j)=Ø
        t=Ø
        sar=Ø
        do 3 j=1,nc
        do 3 i=1,nr
```

```
3       sar=sar+abs(x(i,j))
        s=sar/n
4       do 5 i=1,nr
        rn(i)=0
5       rd(i)=atom
        do 6 j=1,nc
        cn(j)=0
6       cd(j)=atom
        do 7 j=1,nc
        do 7 i=1,nr
        z=x(i,j)-t-re(i)-ce(j)
        u=amin1(1.0,(z/(cutoff*s))**2)
        w=(1-u)**2
        rn(i)=rn(i)+z*w
        rd(i)=rd(i)+w
        cn(j)=cn(j)+z*w
7       cd(j)=cd(j)+w
        rs=0
        do 8 i=1,nr
        re(i)=re(i)+rn(i)/rd(i)
8       rs=rs+re(i)
        cs=0
        tn=0
        td=0
        do 9 j=1,nc
        ce(j)=ce(j)+cn(j)/cd(j)
        tn=tn+cn(j)
        td=td+cd(j)
9       cs=cs+ce(j)
        rs=rs/nr
        do 10 i=1,nr
10      re(i)=re(i)-rs
        cs=cs/nc
        do 11 j=1,nc
11      ce(j)=ce(j)-cs
        t=t+rs+cs-tn/td
        sar=0
        do 12 i=1,nr
        do 12 j=1,nc
12      sar=sar+abs(x(i,j)-t-re(i)-ce(j))
        if(abs(1-s*n/sar).lt.epsilon) return
        s=sar/n
        goto 4
        end
```

INDEX